Security, Development and Human Rights in East Asia

Series Editor
Brendan M. Howe
Graduate School of International Studies
Ewha Womans University
Seoul
Korea (Republic of)

This series focuses on the indissoluble links uniting security, development and human rights as the three pillars of the UN, and the foundation of global governance. It takes into account how rising Asia has dramatically impacted the three pillars at the national, international and global levels of governance, but redirects attention, in this most Westphalian of regions, to human-centered considerations. Projects submitted for inclusion in the series should therefore address the nexus or intersection of two or more of the pillars at the level of national or international governance, but with a focus on vulnerable individuals and groups. The series targets postgraduate students, lecturers, researchers and practitioners of development studies, international relations, Asian studies, human rights and international organizations.

More information about this series at
http://www.springer.com/series/14488

Miguel Ángel Pérez Martín

Security and Human Right to Water in Central Asia

palgrave
macmillan

Miguel Ángel Pérez Martín
Universidad Complutense de Madrid
Madrid, Spain

Security, Development and Human Rights in East Asia
ISBN 978-1-137-54004-1   ISBN 978-1-137-54005-8 (eBook)
DOI 10.1057/978-1-137-54005-8

Library of Congress Control Number: 2016957408

Cover illustration: Pattern adapted from an Indian cotton print produced in the 19th century

Printed on acid-free paper

This Palgrave Pivot imprint is published by Springer Nature
The registered company is Nature America Inc.
The registered company address is: 1 New York Plaza, New York, NY 10004, U.S.A.

*Don't judge each day by the harvest you reap but by the seeds
that you plant*
Robert Louis Stevenson

CONTENTS

LIST OF FIGURES

LIST OF TABLES

CHAPTER 1

Introduction

Abstract This chapter comprises an initial study to introduce the reader to both theoretical (human security) and methodological (constructivism) aspects and their relation with the human right to water and the international security organizations in Central Asia, i.e. the Collective Security Treaty Organization (CSTO), Shanghai Cooperation Organisation (SCO), Organization for Security and Co-operation in Europe (OSCE) and various NATO-sponsored programs. Here, the author explores human security from the perspective of the Canadian School, which sees political decisions as being basic to the allocation of water resources, and considers water as a welfare resource. From a methodological perspective, analysis is based on constructivist school and Thierry Balzacq's definition of securitization (2011). Accordingly, "securitization" is understood as a discursive strategic process built on a natural resource within the specific framework of international security organizations in Central Asia.

Keywords Human security · Right to water and sanitation · Desecuritizing · Securitizing · Extraordinary politics · Exceptional politics · Empowerment · Emancipation

Access to water and sanitation are certainly a precondition for the enjoyment of fundamental human rights such as the right to life, health, work, education and protection against torture, as well as being crucial to

© The Author(s) 2017   1
M.Á. Pérez Martín, *Security and Human Right to Water in Central
Asia*, Security, Development and Human Rights in East Asia,
DOI 10.1057/978-1-137-54005-8_1

achieving gender equality. However, its recognition as a human right and the basis of other rights is very recent. In November 2002, the Committee on Economic, Social and Cultural Rights (CESCR) of the UN explicitly recognized access to safe water as a basic human right, and in July 2010 the United Nations General Assembly, through the initiative of Bolivia and 33 other states, recognized access to clean water and sanitation as "basic human rights" (United Nations 2010, p.2). It also called on the international community to facilitate access to water for nearly 900 million people who do not have such access. The human right to water proclaims that "states must guarantee the right of everyone to sufficient, safe, acceptable, physically accessible and affordable water for personal and domestic uses". It also highlights the priority of the production of food in order to avoid hunger and diseases associated with bad water quality. The Declaration on the Human Right to Water and Sanitation (HRWS) implies the inclusion in international practice of certain values, rules and regulations in the current models of water resources management. The implementation of the human right to water is a new set of obligations and rights, not only in the domestic sphere but also in foreign affairs, and its application can have huge geographical implications because most of the water resources that exist across the planet are transboundary. More than 30 countries are located entirely in transboundary basin territories, and in another 39 countries, with a total population of 800 million people, at least half of their water resources come from outside their borders (FAO; Human Development Report 2006, p. 205). The transboundary basins comprise nearly 50 % of the land area, are home to 40 % of the world population and generate nearly 60 % of available fresh water (Giordano and Wolf 2003). The management of water resources in the international basins has become a matter of international security. With regard to the field of international relations, the General Comment included in the human right to water establishes the following guidelines:

1. The activities related to management of water resources within the jurisdiction of a state party should not deprive another country of the ability to ensure that persons within their jurisdiction can exercise that right.
2. The states parties should refrain from imposing embargoes or similar measures that prevent the supply of water. The water should not be used as a weapon power tool.

3. The states parties should take steps to prevent their own citizens and companies from violating the right to water of individuals and communities in other countries.

4. Depending on the availability of resources, States Parties shall facilitate the realization of the right to water in other countries; for example, water resources and providing financial and technical assistance, and providing necessary assistance requested.

5. The States Parties shall ensure that international agreements are given due attention to the right to water and to this end should consider the creation of new international laws.

6. The states parties must ensure that their actions as members of international organizations take due account of the human right to water. Accordingly, States Parties which are members of international financial institutions such as the International Monetary Fund, the World Bank and regional development banks, should take steps to ensure that their lending policies in credit agreements and other international measures take into account the right to water.

From these guidelines we can draw certain conclusions:

(a) The states should be the most responsible guarantors of the human right to water in its territory, respecting, implementing it and punishing domestic or foreign actors preventing or limiting the enjoyment of this right to any sector or community.

(b) Cross-border cooperation between states is a key element to ensure the human right to water. Therefore, states must ensure not only the human right to water in its own territory but also in the territories of neighbouring countries implementing policies of cooperation.

(c) The water should not be used by states as a weapon of political or economic pressure against other states.

(d) Many actors should be involved in the transboundary management of water resources: states, international organizations, user associations, civil society.

(e) Agreements and treaties, signed by the states (especially those related to economic cooperation), must comply or cooperate to ensure the human right to water.

(f) The states should consider the possibility of creating new institutions and international organizations involved in the management of water resources and the human right to water.

The acceptance and implementation by states of these guidelines directly affects the management policies of the states and their international policies regarding water resources. Consequently, the implementation of the human right to water will be determined largely by the concepts of security or security models that are used in the management of water, and how these models conceive water will predetermine the results

WATER, SECURITY AND HUMAN SECURITY

The first ideas about water and security can be found in a number of papers published in the late 1970s and during the 1980s, including "Redefining security" by L. Brown (1977), "Redefining security" by R. Ullman (1983), and "An Expanded Concept of International Security" by A. H. Westing (1986). The main conclusion of this work was the claim that most of the conflicts between states are rooted in competition for territory or natural resources (including water). If natural resources are in decline, conflicts tend to increase. In many cases the scarcity of natural resources is connected to the degradation of the environment. Therefore, the environment, water resources and security are inextricably linked. During the 1990s authors such as Homer-Dixon (1991) and Gleick (1992; 1993) began to create the first classifications of water conflicts. Homer-Dixon, in his "On the Threshold: Environmental Changes as Causes of Acute Conflict. Part 1" identified three types of conflicts resulting from environmental changes:

1. **Simple Scarcity Conflicts**. Simple scarcity conflicts are explained and predicted by general structural theories. They are the conflicts we would expect when state actors rationally calculate their interests in a zero-sum or negative-sum situation such as might arise from resource scarcity.
2. **Group-identity Conflicts**. Group-identity conflicts are explained and predicted by group-identity theories. Such conflicts are likely to arise from the large-scale movements of populations brought about by environmental change.
3. **Relative-deprivation Conflicts**. Relative-deprivation theories indicate that as long as developing societies produce less wealth because of environmental problems, their citizens will probably become increasingly discontented by the widening gap between their actual level of economic achievement and the level they feel they deserve.

Definitively, in Homer-Dixon's opinion, conflicts about water are determined primarily by the scarcity of resources, motivated by environmental anomalies. Meanwhile, Gleick, in his article "Water and Conflict", gives us a wider and essentially political vision of water conflicts. The threats can be grouped into five categories:

- **Control of Water Resources**: where water supplies or access to water is at the root of tensions.
- **Military Tool**: where water resources, or water systems, are used by a nation or state as a weapon during a military action.
- **Political Tool**: where water resources, or water systems, are used by a nation, state, or non-state actor for a political goal.
- **Terrorism**: where water resources, or water systems, are either targets or tools of violence or coercion by non-state actors.
- **Military Target**: where water resource systems are targets of military actions by nations or states.
- **Development Disputes**: where water resources or water systems are a major source of contention and dispute in the context of economic and social development.

Gleick essentially considers water as a power resource. Water is used as a power lever, and this situation is the main reason of water conflicts. Let us analyze the most recent last water conflicts classification, the Transboundary Freshwater Dispute Database, which includes a computer compilation of 400 water-related treaties, negotiating notes and background material on fourteen case studies of conflict resolution, news files on cases of acute water-related conflict, and assessments of indigenous/traditional methods of water conflict resolution. This database was partially developed by A. T. Wolf (Professor, Department of Geosciences, State University of Oregon, USA). This author establishes a direct link between water resources and human security. In his article "Water and human security" (Wolf 1999) he stated that "The greatest human security threat of the global water crisis comes about not from the threat of warfare, or even from political instability, but rather from the simple fact that millions of people lack access to sufficient quantities of this critical resource at sufficient quality for their wellbeing." The real cause of water conflict is not its scarcity or its use as a tool of power, but the lack of access to water resources. In essence, water conflict is a problem of mismanagement of water resources. According to A. T. Wolf's thinking, we can also say that The resolution of water conflicts passes for establishing a universal access to these resources as proposed by the human right to water and sanitation."

In order to establish the existing relationships between security, water resources, multilateral organizations and the human right to water, in this investigation I shall use the security model or theoretical framework on human security from the Canadian School perspective (J. Nef 1999; Alkire 2003), that sees political decisions as basic to the allocation of water resources, and considers water as a welfare resource. And, from a methodological perspective, I shall use constructivism, focusing precisely on how those decisions are generated (international organizations). Several reasons have encouraged me to use human security, HRWS and constructivism in my investigation, some of them theoretical, and some others of a methodological or ethical character: human security offers an excellent model to develop the human right to water. Some reason explains this point:

1. Human security (CP), as well as HRWS, whose policies prioritize the security of individuals and different aspects that are basic to enabling people's development and dignity (Nef 1999),[1] unlike other security approaches focused more on border defense, sovereignty or state vital centers.

2. Water is a resource involved in all aspects of human life, and human security offers us a study agenda related to every dimension including this resource (security, politics, economy, food, ecology, health, individuals, community) and, thanks to some indicators, we can know the importance of this resource in each of its aspects and how the security of individuals is affected by difficult access or the complete lack of access.

3. Water is a liquid element in permanent motion, not usually constrained by physical or political borders. Cooperation between two or more states thus becomes essential for an ideal exploitation of this resource (Wolf 1999). Human security is based, among other things, on cooperation as its main tool in order to avoid conflicts and to create organizations and institutions to prevent them, as suggested by the HRWS.

4. Water is often a variable resource in terms of distribution, volume and quality; so that provision, prevention and planning of water resources use and management are all indispensable factors for the availability of a safe, continuous, sufficient and equitable supply among its users (Wolf 1999). Human security considers provision, prevention, equitability and a sustainable management or development of this resource as indispensable elements to avoid risks or conflicts and to guarantee the HRWS.

5. Shortage of water or water unsafe to drink is not always the cause of an armed conflict, but it has a strong impact on health, quality of life and the development chances of individuals and their communities. Human security does not assume that a conflict must necessarily become violent (Nef 1999); in this way it helps in gauging the dimensions of the conflicts in all their aspects and preventing them from damaging the HRWS.

6. There are many actors involved in the management of water resources. Human security considers that all the actors who take part in this process must participate with no distinction, so that consensus prevails over disagreement in water resources management, whose ultimate aim must be to see that access to water and sanitation as a human right is more important than particular interests, because according to the human security values, the object to be guaranteed must be the human rights, both individual and collective (Alkire 2003).

7. The Canadian perspective of human security lays particular emphasis on the relevance of political factors in the decision-making process, and opens the door to civil society's participation and transparency in information (Alkire 2003), both of these being fundamental to HRWS implementation and protection.

8. HRWS, in the General Comment No. 15, underlines that water must be considered as a social and cultural good, not only as an economic one, and that the right to water must be exercised in a sustainable way. Human security has at the core of its theory the protection of the individual against critical threats (Alkire 2003). As a consequence, it conceives water as a welfare resource aiming to protect individuals' security in all its dimensions (political, economic, social, etc.).

These are some of the rules and values provided by the approach of human security (CP) in the water management resources field, with the purpose of guaranteeing the HRWS and preventing or avoiding conflicts between countries.

Human Right to Water, Constructivism and International Organizations

Security is seen by constructivism as a concept elaborated independently of the actors to be "protected," a product of the actors' interaction in a specific context (Lipschutz 1995, p. 10). In short, security has a high degree of subjectivity and will be whatever actors want it to be (Wendt 1992). It is obvious that this flexible concept of security eases the inclusion of new values and practices in security models such as the HRWS, but including water resources and their management within the security studies sphere is basically owed to the School of Copenhagen, composed of B. Buzan, O. Waever and J. De Wilde. These authors, in the book "Security: a new context for its analysis" (Buzan et al. 1998), argued that environmental questions must be included in international security agendas as relevantly as other aspects: military, political, economic or social, because all these dimensions are interconnected in some way. After including the environmental dimension, water should automatically be incorporated into the new twenty-first-century security agendas. According to Buzan, water was an element destined to suffer a securitization, or desecuritization, process, so that it should be included in the security complexes defined by the author as "a set of units whose main securitization and desecuritization processes are intertwined in such a way that their security problems cannot be analyzed nor solved separately". These ideas thus allow us to say that cross-border watercourses make up a security complex in which states' security is an interdependent concept. In these cross-border watercourses, water resources management must be shared by "securitizing" or "desecuritizing" water-related policies to guarantee the HRWS. I shall use another relevant aspect of the constructivist methodology focusing on some essential actors in building interstate relationships, such as multilateral organizations connected, in this case, to security policies.

To carry out this research I shall start from Balzacq's definition of securitization: "securitization as an articulated assemblage of practices whereby heuristic artefacts are contextually mobilized by a securitizing actor, who works to prompt an audience to build a coherent network of implications about the critical vulnerability of a referent object, that concurs with the securitizing actor's reasons for choices and actions, by investing the referent subject with such an aura of unprecedented threatening complexion that a customized policy must be undertaken immediately to block its development."

In short, in this research securitization is understood as a discursive strategic process built on a natural resource, water in this case, within the specific framework of international security organizations. The building of this discourse can lead into several different stages, such as "normal politics or desecuritization," "extraordinary politics" and "exceptional politics."

THE NORMAL POLITICS OR DESECURITIZATION

This process occurs when "the issues are moved out of the security sphere and back into the political one, which has narrower and more mundane competition features due to diverse interests and the elite political management" (Kalyvas 2008, p. 164).

THE EXTRAORDINARY POLITICS

It stresses the possibility of securitization as a process of openness and self-determination with democratic potential (Williams 2015, p. 115). The democratic politics of extraordinary refers to those infrequent and unusual moments when the citizenry, overflowing the formal borders of institutionalized politics, reflectively aims at the modification of the central political, symbolic and constitutional principles and at the redefinition of contents and purposes of a community (Kalyvas 2008, p. 7).

THE EXCEPTIONAL POLITICS

It produces an exclusionary order, the logic of exception postulates that security is about the fight against an existential threat that necessitates "exceptional measures" (Schmitt 2005, pp. 13–14), and focuses on speech acts legitimizing exceptional policies and practices in the face of an existential security threat.

The achievement of this goal requires the analysis of the nature of water securitization processes. According to Balzacq (2008), we need to identify at least three sets of factors:

A. Audience; the recipients of the message when
   (a) it has a direct causal connection with the issue; and
   (b) it has the ability to enable the securitizing actor

to adopt measures in order to tackle the threat. Summing up, securitization is satisfied by the acceptance by the empowering audiences of a securitizing move. The audience will accept the discourse when the securitizing actor has the ability to identify the audience's feelings, needs and interests.

B. Context and codependency agency; in the words of Balzacq (2010, p. 64), "the critical question is not whether discourse 'does' things, but instead under what conditions the social content and meaning of security produces threats…We will analyze in which political, environmental, social or economic contexts the securitization processes on water appear"

C. The dispositif and the structuring force of practices; the dispositif is defined by Balzacq (2008, p. 79) as "an identifiable social and technical 'dispositif' embodying a specific threat image through which public action is configured to address a security issue". According to Balzacq, these dispositifs have some particular features:

1. Defining features that align with others and design traits that make it unique.
2. They are tools that configure actions, procedures, skills requirements, a routinized set of rules, etc. and create a structure for the relations between individuals and institutions.
3. The tools of securitization reconfigure and call public attention to what is seen as a threat. They also create a specific image of the essence of that threat.
4. The dispositifs essentially use two kinds of tools;

   A. Regulatory instruments that provide the framework within which capacity tools operate.
   B. Capacity tools; the dispositif generates a series of capacities that allow the making of decisions and carrying out of activities.

Therefore, in this research we shall apply this model of analysis focused on the study of international security organizations and their water policies.

MULTILATERAL ORGANIZATIONS

The choice of international security organizations in this chapter as the priority context of analyses is based on the constructivism of Alexander Wendt.

According to Wendt (1992, p. 400), the international system is constructed from social structures in which the ideas and symbolic interactions define how states relate to each other. Consequently, system structures are socially constructed, and the interests and identities of states are conditioned by social structures and vice versa. International security organizations as agents of the social construction of reality select events or issues that are considered critical for coexistence between states. These organizations have a great ability to influence the international agenda-setting states, as well as being core binders of all kinds of resources—resources that are required to build common strategies for development of the human right to water and avoid conflicts or neutralize threats that transcend state borders, such as conflicts associated with water management.

The scope of this statement is long and complex, and concerns international organizations, treaties and international agreements. The international institutions involved include issues related to natural resources, environment, economy, politics and security, as water is an irreplaceable resource, present in all human aspects, and it does not recognize political boundaries and it is distributed unevenly among different regions.

The policies and strategies of multilateral security organizations in the protection of this new human right as part of the maintenance of peace between states is one of the most important debates in this area. The implications are many, including multilateral cooperation regarding water resources and the responsibilities of the international security organizations in this regard. A privileged setting for this analysis is the Aral Sea basin in Central Asia, and more specifically, the Amu Darya river, an international waterway shared by Tajikistan, Afghanistan, Turkmenistan and Uzbekistan. Currently there are conflicts related to water management, particularly among the countries situated at the head of the river, such as as Tajikistan and Afghanistan, and those found downstream, such as Uzbekistan and Turkmenistan, in situations of dependency.

While states are the main actors, multilateral organizations, from the constructivist standpoint, play an important part in international dynamics because some of them can establish rules that are able to change states' policies: international organizations have a great potentiality to generate fundamental and regulatory rules regarding a specific matter or area. Some international organizations act as the contact point where international actors learn to cooperate in setting up a common identity.

The cooperation relationships axis or guiding thread in questions of security between the great powers and Central Asian countries will be undertaken, as a priority, through regional and international cooperation organizations operating in the area:

- the Collective Security Treaty Organization (CSTO);
- the Shanghai Cooperation Organisation (SCO);
- the North Atlantic Treaty Organization (NATO); and
- the Organization for Security and Co-operation in Europe (OSCE).

The consideration of these organizations through which to study security relationships between Central Asian countries and the great powers is justified by several methodological and theoretical points:

1. It is increasingly difficult to assume that a country's security can be seen exclusively from either a realistic or a selfish perspective. Security cannot depend only on particular interests or the capacity to impose them, but on several countries establishing certain coexistence, trust and action frameworks. Security has thus gradually developed a more collective meaning, and organizations contextualize states' actions by giving meaning to their actions or denying it.
2. Regional or international cooperation organizations arrange and select those events or questions considered as basic to coexistence among states, and need to be solved through a collective approach. International institutions therefore have more influence day to day the states' international agendas.
3. Regional or international cooperation organizations act as public forums where states hold open discussions, establish agreements (or disagreements), make (or do not make) decisions, and set up strategies regarding a specific question. They detail the priorities of their political agenda, the abilities of their members to reach agreement, and the reasons why they do or do not agree.
4. The states build their identities and perspectives in these public forums, not only in relation to other organization members, but also to those that do not belong to it.
5. A study of the organizations allows a more complex and multi-dimensional vision of the matters discussed to be obtained, and thus it will help in the creation of strategies.

In short, security is gradually acquiring a more collective meaning, and states' actions are contextualized by the organizations, either supporting their actions or denying them. In this constructivist perspective, the study of organizations will be carried out under the following premises:

1. Organizations might become even more than merely the reflection of the power balance of states, as realists think, or a cheap tool for cooperation, as the liberal school considers.

2. When straightforward interventions are more expensive than a collective approach, international organizations can accomplish important functions:
   (a) Classifying the world of international relations, creating categories, actors and actions;
   (b) Establishing meanings in the social world; and
   (c) Articulating and spreading new rules and principles.

In short, regional or international cooperating organizations arrange and select those events or questions considered to be basic to the coexistence of states and need to be solved through a collective approach. Regional organizations therefore have an increasing influence over international agendas, policies and attitudes, generating normative and fundamental laws in both local and international spheres.

However, the accomplishment degree of these premises will vary, depending on two factors(i) the autonomy and authority of the international organizations regarding their members, and (ii) some others such as the lack of resources or some faults in the internal functioning of the bureaucratic structures.

According to Wendt (1992, p. 400), institutions and organizations do not need to be eminently cooperative; they can be competitive (states identify themselves negatively for the sake of their own security and ego), individualistic (states are indifferent to connections between their own security and the security of others; they focus only on sharing profits), or cooperative (states identify themselves positively, and security is perceived as a joint responsibility).

Therefore we should wonder in this research whether international security organizations in Central Asia consider the access to water and sanitation as a security matter and cooperation with regard to water resources as a priority.

HRWS, EMPOWERMENT AND EMANCIPATION

One of the key aspects to guarantee HRWS, according to the Comment 15, is the empowerment of both the individual and civil society as a whole. To implement any human right, the right to water included, participation must lead to a true empowerment, not just a simple query or supply of information. An active, free and meaningful participation needs a specific chance to express claims and concerns as well as to influence decisions. Thus it is necessary to provide information through multiple channels, allowing participation in transparent and inclusive processes, and strengthening the collaboration abilities of individuals and civil society within the international security organizations.[2]

However, to guarantee the empowerment of individuals regarding HRWS or any other field of human rights, we must mention a broader concept: emancipation. According to K. Booth (1991), with regards to security, emancipation should have a central position in matters such as power and order, because power and established order are generally favored at the expense of non-power and the instability of others. Emancipation is, in Booth's opinion, a level of security that provides individuals and groups with time, energy and chances to do something more than just survive as human biological organisms.[3]

To implement all the activities proposed by Booth, water is an essential and irreplaceable resource because of its involvement in every aspect of human life (health, hygiene, economy, environment, etc.), a resource that enables the emancipation process which, according to this author, is the basis of real security (Booth 1991, pp. 313–326). HRWS becomes an unavoidable protocol to reach the emancipation proposed by Booth and the maximum level of security implying a universal access to water with transparent, participative, responsible, sustainable and cooperative management in the international sphere, even though it means "a radical change in the distribution of power and in the way that power is conceived and exercised. It involves an attempt to empower the disenfranchised and to give a voice to those that have traditionally been silenced" (Christie 2010, p. 181). To achieve HRWS is a political challenge and not just a simple distribution of a resource that gives individuals or groups the chance to participate freely and unthreatened in the decision-making process.

PARTICIPATION, CIVIL SOCIETY AND MULTILATERAL ORGANIZATIONS

International organizations involved in security cooperation should not only see water as an accessible good to ensure the security of all human beings, but must also guarantee the participation of individuals and civil society groups as citizens in the decision-making process. According to Keith Krause and Michael Williams, for example, "security is synonymous with citizenship" (1997, p. 43), such that creating effective states capable of delivering the goods of citizenship is the *sine qua non* of providing human security.

One of the keys to the empowerment of individuals and communities, and therefore also to reach the aims of HRWS, is the participation, free of fear and threat, of civil society in the decision-making process of multi-lateral security organizations. Because of that, in this research we shall analyze the involvement of groups and members of civil society in the participation and decision-making contexts related to water resources. Through this, we shall see whether those security organizations do or do not empower the individuals or groups to make their rights feasible – in this case the rights to water and sanitation.

Despite the relevance of the regional organizations and their ability to inform and lead us in the international relationships study, it is obvious that not "everything" happens within them or they do not explain "everything;" description and analysis of systems or contexts where the organizations' development takes place are therefore decisive for their understanding. These contexts are composed of the juxtaposition of five sub-systems (Nef 1999):

1. Ecology or environment (context, culture, structure, process and effects).
2. Economy (context, culture, structure, process and effects).
3. Society or socio-demographic system (context, structure, process and effects).
4. Politics (context, structure, process and effects).
5. Culture (context, structure, process and effects).

Each sub-system is structured around a set of relatively homogenous and recognizable questions reflecting the specific nature of its constituent elements (context, structure, process and effects), which give meaning to the actions. In this research we shall therefore describe and analyze those aspects or factors of the international context that have a decisive effect on the different dimensions of human security, the regional security organizations and water resources management in Central Asia.

Notes

1. *Environmental security:* the right of individuals and communities to preserve their lives and health in a safe and sustainable environment. *Economic security:* chances of access to employment and necessary resources for survival as well as decrease of needs and quality of life for the whole community. *Society security:*

eradicate gender, age, ethnic group or social status discrimination within a community, this idea involving free access to knowledge and information networks and the possibility of association. *Cultural security:* a set of psychological orientations adapted to the people's needs to preserve their ability to control uncertainty. *Political security:* the right to be represented, to be free, to participate, to dissent, all these rights combined with the possibility of celebrating elections reasonably and probably leading to real changes. This idea includes certain law and judicial guarantees, individual and collective, as access to justice and protection against abuse.

2. OHCHR. See http://www.ohchr.org/Documents/Issues/Water/LegalObligations_sp.pdf.

3. World security conceived as "more than just surviving" means to create conditions in which the global "we" (and not only the dominant current one) can live a public and personal life with dignity, laughter, music and proper meals.

CHAPTER 2

A Water History of Central Asia

Abstract The management of water resources in Central Asia in the second half of the twentieth century was catastrophic, its best example being desiccation of the Aral Sea in a region where an average of 30 % of the population has no access to drinking water. Moreover, water management is the origin of frequent conflicts and tensions between countries with a surplus of water resources (Tajikistan, Kyrgyzstan, Afghanistan, for example) and those with a strong degree of dependence on water coming from them (such as Uzbekistan and Turkmenistan). Nevertheless, Central Asia's disputes over fresh water are not just a contemporary question but have a long and distressing history. In this chapter, the author conducts a historical and geographical study on water in the region and how the use and management of this resource has shaped political and social events since the nineteenth century.

Keywords Treaties · Colonialism · Frontiers · Ethnic groups · Soviet Union · Independence · Water control

The history of international treaties concerning water goes back to 2,500 BC, when the two Sumerian state-cities of Lagash and Umma established an agreement to put an end to a fight about the Tigris river's water resources use—this is considered to be the first known international treaty in history.[1]

© The Author(s) 2017   17
M.Á. Pérez Martín, *Security and Human Right to Water in Central Asia*, Security, Development and Human Rights in East Asia,
DOI 10.1057/978-1-137-54005-8_2

There have been a considerable number of treaties about water since then. According to the Food and Agriculture Organization of the United Nations (FAO), more than 3,600 treaties related to international water-courses have been written since 1985.[2]

Most of these are related to ship navigation and demarcation of limits. However, during the twentieth century the international community shifted its interest from navigation to the management, use and preservation of water resources. This new motivation or focus on water resources management and governance has increased since the early 2000s, for a number of reasons:

1. The dissolution of the Soviet Union and the collapse of the bipolar international system has had as a consequence encouraged the birth of new concepts regarding security in academic spheres, foreseeing new kinds of risks or threats, such as the environmental crises connected to the shortage or mismanagement of water resources.
2. Concern among the scientific community regarding the consequences of climate change on water resources.
3. The claims of international organizations and civic associations about increasing levels of pollution in rivers and seas, caused by non-sustainable models of economic and industrial development that spoil the environment, water and the citizens' quality of life.
4. Water privatization processes backed by the World Bank (WB), the International Monetary Fund (IMF), the World Trade Organization (WTO), companies and other organizations, have sometimes produced important civil conflicts, as in the case of Bolivia.
5. The persistence of hunger in the world, as well as deprivations related to water resources: according to the UN, about 884 million people have no access to drinking water and more than 2,600 million have no basic sanitation facilities; approximately 1.5 million children under 5 die every year because of water- and sanitation-related diseases. As a consequence, in August 2010, the UN declared the right to drinking water and sanitation to be a human right (HRWS) essential for the full enjoyment of life and all human rights.

Item 5 here is especially relevant because the Declaration of Access to Water and Sanitation as a human right strengthens the international component and the importance of cooperation in cross-border regions in matters of water resources, for several reasons:

1. A great part of the water resources existing in the planet are located within cross-border regions. The cross-border basins include almost 50 % of the planet's surface, 40 % of the world's population, and nearly 60 % of the available fresh water.[3]
2. More than 30 countries have their whole territory placed within cross-border basins. Thirty-nine countries, and their 800 million inhabitants, have at least 50 % of their water resources coming from beyond their borders.[4]
3. Because of the condition of water as a non-replaceable, ever-changing, irregularly distributed and mobile resource, strategically involved in all aspects of human life (personal use, hygiene/health, agriculture, industry, culture), the use, access or management related to water are not just a technical problem with engineering implications,[5] but also a political process and, because of its transnational nature, this resource also in many cases raises questions of international security.[6]

Many countries, the Central Asian ones among them, consequently depend on the use of cross-border watercourses for their socioeconomic development, as well as to guarantee the human rights of their population.

Water has been, and remains, a vital element for human development to guarantee human rights. The lack of access to it, its poor quality or restrictions of use, resulting from either natural or human causes, have become some of the most fearsome threats to human wellbeing and have originated domestic and international conflicts, directly or indirectly, throughout history, as well as the violation of millions of people's human rights. This idea implies that conflicts derived from the management of water resources have a direct effect on the security of states, the wellbeing of their populations and world peace in general.

However, the answer from the international community in situations of conflict between countries sharing the same watercourse, or in violations of the human right to water, has been ambiguous, not very convincing, and it has not allowed an international consensus strong enough to cope with problems derived from water management in a unanimous way.

Several international agreements and forums—none of them legally binding—have attempted to solve some conflicts between countries in the matter of water resources management; for example, Convention on the Law of the Non-Navigational Uses of International Watercourses (1997) or the Dublin Principles (1992). This lack of commitment and unanimity between states, in order to establish a few basic universal principles about the management of water resources, also affects the consideration of access to water and sanitation as a human right, the very foundation of all other rights.

In the voting session to approve the resolution on the human right to water, 122 countries were in favor of it, 41 abstained and 29 were absent. In Central Asia, where the cross-border resources management is the reason for bitter disputes and millions of people have not guarantee of their right to water and sanitation, the voting outcome was: Tajikistan, Afghanistan and Kyrgyzstan voted in favor of the resolution; Kazakhstan abstained; and Uzbekistan and Turkmenistan, the main water consumers and less able to renovate their water resources, were not present at the voting session.

The management of water resources in Central Asia in the second half of the twentieth century was catastrophic and the worst example is the desiccation of the Aral Sea, though this is not the only problem in a region where now about 30 % of the population have no access to drinking water, and in some areas this proportion goes up to 50 %.[7] Moreover, water management is currently the origin of frequent conflicts and tensions between countries with a surplus of water resources (Tajikistan, Kyrgyzstan, Afghanistan) and those with a strong degree of dependence on water coming from them (Uzbekistan and Turkmenistan). Conflicts here are basically over two cross-border watercourses in the region—Amu Darya and Syr Darya—and this research will focus on the Amu Darya river.

WHY THE AMU DARYA?

The Amu Darya river is the most important water reserve in the region, its basin is the most populated and water resource cooperation is more precarious than in the Syr Darya basin, the other great river in the Aral basin. Therefore, the conflicts, risks and threats might become worse and bigger in this watercourse.[8]

The human rights and security of the countries in the Amu Darya basin are in a serious crisis, produced precisely by non-sustainable water resources management related to the poor degree of cooperation between countries in this matter, their lack of commitment to guarantee the human right to water, and the need of an international binding legislation for water resources management in a context where global warming, a growing demographic, increasing river basin pollution and greater water demands, for both agriculture and industry, will make water—and not only in Central Asia—a good gradually more and more appreciated.

Various multilateral security organizations have since the 1990s been developing strategies in the region to cope with the conflicts and threats that are devastating this environment. Because of the seriousness of the current conflicts related to water, there is room for some questions as to whether these organizations have considered the conflicts and the threats connected to water as their own problem, whether they have offered Central Asian countries all the necessary incentives to cooperate in matters of water resources, and whether they have in their agenda the human right to water and sanitation, with the purpose of solving the troubles related to water resources management and guaranteeing their populations the most basic of all human rights: the right to water and sanitation (HRWS).

In this region of the world, the HRWS might become the true key in terms of security—the guarantee of the wellbeing, human rights and peace in its societies. Therefore the HRWS start-up will be determined mainly by the concept or models of security used in water management and the way these models conceive that water will predetermine the outcome obtained. Whether water is managed as a natural resource susceptible to be used as a power resource (Falkenmark and Lindh 1974; Glieck 1993); as an item of merchandise or a production factor (Winpenny 1994; Rogers 2002; Homer-Dixon 1991); or as a welfare resource (Wolf 1999), the outcome will be different, as well as its consequences for the users and agents involved in the process.

THE SCENE

The general geographic framework upon which this investigation is focused is the territory covering the five former Soviet republics in Central Asia: Kazakhstan, Turkmenistan, Kyrgyzstan, Tajikistan and Uzbekistan, and, more specifically, the Amu Darya basin, as well as, on occasion, Afghanistan and its relations with its northern neighbors, considering its historical, political and economic connections with Central Asia and the importance of this country as a producer of water resources in the Amu Darya basin.

GEOGRAPHY OF CENTRAL ASIA

The territory of Central Asia is located between Europe and Asia and covers up to 4 million square kilometers, about half the area of continental USA (excluding Alaska). The distance from north to south is 2,413 km, and 3,128 km from west to east. Its geographic limits are the Caspian Sea in the west, the Altai and Tien Shan mountain ranges in the east, the

Siberian taiga in the north, and the Kopet Dag, Pamir and Hindu Kush mountains in the south. This huge area contains several river basins; the most important of these being the Syr Darya, Amu Darya, Ural Emba, Chu, Ishin, Tobol and Irtysh. The countries included in this territory are Kazakhstan, Uzbekistan, Turkmenistan, Tajikistan and Kyrgyzstan. Its political limits are the Russian Federation to the north, China to the east and Afghanistan and Iran to the south.

CENTRAL ASIA, AFGHANISTAN AND THE AMU DARYA: A CONTROVERSIAL BORDER

Central Asia has traditionally been a huge border space between several great civilizations: Turkish-Iranian, Mongol, Chinese, Slavic and Hindu. All these cultures converged around Central Asia, and the region, in spite of its complex geography and adverse climatic conditions, became a bridge for transcontinental communication. Its strategic spatial situation, between Europe and Asia, has been one of the most influential variables in its long history. The Silk Road, one of the first and most important of the inter-continental trade routes, connecting the Mediterranean with China and transporting all kinds of goods and ideas, had in the Central Asian oasis some crucial spots for its development. In the eighteenth century, the Russians, as had the Mongols a few centuries before in order to fulfill their ambitions for new lands, riches and power, used this region to expand their empire to the boundaries of India and China (Fig. 2.1).

Before the Russia's annexation and conquest of the Central Asian territories, the inhabitants of the region were divided in two sociopolitical and economic organizations: khanates and hordes (nomads).[9]

The khanates of Kokand, Bujara and Khiva, located within the Syr Darya and Amu Darya basins and in the Fergana Valley, were inhabited by a mixture of Iranian peoples (mainly Tajiks) and Turkish-Mongolians (mainly Uzbeks). The economies of these khanates were based mainly on intensive agriculture, enabled by a complex network of waterways swallowing the surplus production necessary for survival and trade, though artisans and traders also played an important role. The structure of power in the seden-tary oasis society was strongly hierarchized: at the top was the sultan or the khan, backed by an administrative apparatus led by a vizier or cadi, who managed the life of the khanate. The sultan acted as a mediator between the main political actors in the khanate, the khans (largest landowners), the

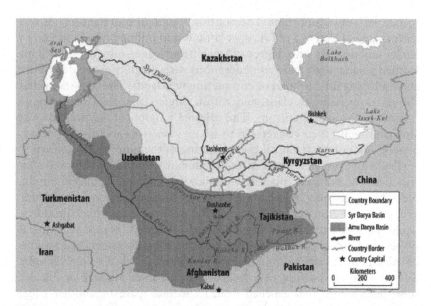

Fig. 2.1 Course of the Amu Darya river and its boundaries

"Avoiding Water Wars: Water Scarcity And Central Asia's Growing Importance for Stability in Afghanistan and Pakistan". A Majority Staff Report Prepared for the Use of the Committee on Foreign Relations United States Senate, One Hundred Twelfth Congress. First Session, February 22, 2011. Printed for the use of the Committee on Foreign Relations. Available via World Wide Web: http://www.foreign.senate. gov/imo/media/doc/Senate%20Print%20112–10%20Avoiding%20Water%20Wars%20Water% 20Scarcity%20and%20Central%20Asia%20Afgahnistan%20and%20Pakistan.pdf.
Source: CRS produced using U.S. Department of State, International Land Boundaries, https://www. intelink.gov/basestate/landBHome.asp; U.S. Geological Survey, HydroSHEDS, http://hydrosheds. cr.usgs.gov; World Resources Institute, Watersheds of the World, ESRI Data and Maps 9.3.1; DeLorme World Vector Data, 1:250,000; IHS World Data, December 2008.

clergy and Sufi orders, either allies of the sovereign or conspirators against him. The legal framework of the khanate was ruled by the Sharia and the Adat (custom), and justice was usually administered by clergymen.[10]

The tribal confederations were composed of Kazakhs, Kyrgyzs and Turkmens. Kazakhs and Kyrgyzs lived in the large territories of the steppes to the north-west of the Caspian Sea, and the Turkmens in the deserts located on the south-eastern coastal areas of the same sea (currently Turkmenistan). Their economies were based on nomadic pastoralism, with sheep, yaks, goats, horses, etc. Nomads needed large areas of land on which to practice transhumance to provide grazing for their livestock. Trade (slaves), looting and tributes received for protecting trade routes were also important. The political foundation of the nomads was the organized clan as a political, economic and military unit guided by a leader sanctioned by his degree of kinship and leadership qualities. Clans formed alliances to create confederations with the purpose of defending themselves against a threat or conquering new lands. The laws that ruled the relations between clans, and within single clans, were orally transmitted tradition and custom. The original religion of the nomads was shamanism and their conversion to Islam was a slow and discontinuous process not completed until the beginning of the nineteenth century.[11]

Trade and war alternated as the predominant relations between both societies. The nomadic tribes became the oasis peoples' suppliers: of milk, meat, hides, clothing, rugs and slaves (either Russians or from other ethnic groups captured in *razzias* [hostile raids]) were their main trade products, while the sedentary tribes exported cereals, weapons, horses' tack, metal objects and luxury goods. Violent periods were usually caused by invasions of other peoples coming from peripheral regions, or when a nomadic tribe lost its livestock because of a lack of water, natural disaster or disease; sometimes a tribe was expelled from its natural space by another tribe, and the nomads invaded and looted nearby oasis towns.[12]

SOCIETY AND WATER RESOURCES

Water resources were extraordinarily important for both the nomadic and sedentary societies. Nomadic societies, highly dependent on their livestock, needed to know of or establish a wide network of watering troughs. One of the main sources of legitimation for the Kazakh leaders (khans) was the knowledge of these routes, their ability to travel them with their cattle herds and to find springs, not only to obtain water supplies but also to locate grazing areas.[13] On the other hand, the economy of sedentary societies was based mainly on irrigated agriculture and they had developed sophisticated models of water management.[14] These two water production and management systems were gradually transformed after the Russian conquest.

THE RUSSIAN CONQUEST

After more than a century of military efforts, the Russian empire moved its border from the north of the Kazakh steppes to the Amu Darya shores.[15] Long distances, a hard, dry climate and a hostile population made the advance of any army through the Central Asian steppes and deserts at that time a difficult challenge: in fact, the first stages of the Russian expansion in the region ended in disaster.[16] Central Asian rivers, lakes and seas became alternative supply channels for the Russian armies during the conquest. In 1848, A. I. Butakov did the first hydrographic study of the Aral sea and a few years later built the first war flotilla, with the purpose of going upstream on the rivers of Central Asia and supporting the military expansion by land of the Russian armies into to the heart of the region.[17]

The Caspian Sea was a very important means of penetration into the region for the Tsar's armies. In 1869, the Russians established a military base named Krasnovodsk on the eastern shore of the Caspian Sea (a town now known as Turkmenbashi), from where the railway's works began, making possible a rapid economic, political and military integration of this area into the Russian Empire.[18]

Central Asian rivers and seas were not only used by the Russians as a way of penetrating the territory: the main waterways and channels that supplied water to cities such as Tashkent, Jizzakh or Samarkand were either altered or cut off, acting as another weapon to put an end to the resistance of these urban centers.[19]

Control over Central Asian rivers and seas, and their logistics benefits, made the conquest easier for the Russians, while the indigenous population's gradual lack of power with regard to the access to and management of water resources hastened their loss of autonomy as well as increasing their dependence on their new Russian rulers.

THE KAZAKH STEPPES

The confinement of Kazakh hordes in small districts—called *volost* by the Russian authorities—the suppression of the great migratory routes, the control the Russian military patrols exercised over the movements of the Kazakh groups of people[20] and, finally, the process leading to a sedentary lifestyle that began in the Soviet era, together meant the end of the Kazakh tribe's control and knowledge of the environment, as well as their freedom to obtain the necessary water and grazing for their cattle herds.

THE OASIS

The khanates located on the Amu Darya and Syr Darya shores lost the control and management of their water resources during two clearly different stages: in imperial years there was an attempt to put the old administrative hierarchies in command of water management (Mirab, Ariq-Aqsaqals, etc.) under the control of the Russian tsarist army and administration[21]; and later, during the Soviet era, a new network of channels and dams was built and managed by technicians coming from the metropolis, serving Moscow's interests, putting completely aside the old water management administrative hierarchies and, at the same time, their values, interests and technical knowledge.

THE EMPIRE IN CENTRAL ASIA

The Russian Empire, after more than a century of military efforts, managed to move its northern border from the Kazakh steppes to the Amu Darya shores.[22] The sultanate of Bujara and the khanate of Khiva were legally considered to be sovereign states, but their independence from Russia was just a formality and their southern border would become de facto the Russian Empire's frontier. The Tsar's armies built a range of checkpoints along the Amu Darya and created the first war flotilla to patrol its shores, to monitor the traffic of people and merchandise across the watercourse. Meanwhile, the Russian Crown acted as guardian to the foreign affairs of the khanates.[23]

THE AMU DARYA: A NEW FRONTIER

The Russian advance across Central Asia meant the integration of the region within a new geopolitical world context. The shifting of the Russian conquest's center of gravity from the Kazakh steppes to the Amu Darya was felt to be a direct threat by other great empires (Persia, China and Great Britain). The arrival of Russian troops on shores of the Amu Darya caused several important repercussions in neighboring areas:

– *South of Amu Darya.* A series of war episodes began; these are known as the Anglo-Afghan wars (1839–1842), (1878–1880) and (1919); their common cause is the loss of confidence of the British Crown in the successive khans ruling the territories that separated British dominions in India from the Russian advance. The British always suspected an alliance between the Afghan tribes and Russia, an alliance that became one of their worst nightmares and the largest threat to their possessions in India. The British Crown thought of warfare as the only means to ensure the Afghan tribes' loyalty, even though only temporarily.[24]

– *West of Amu Darya, Persia, the Caspian Sea and the Caucasus.* The Russian pressure from the north and that of the British from the south put an end to traditional Persian aspirations of hegemony over the Caucasus and Central Asia, regions that were for centuries under their political and cultural influence. After two consecutive war episodes (1804–1812) and (1827–1828), and their respective treaties, Russia secured its imperial rule over Southern Caucasus (Georgia, Azerbaijan and Armenia) and its fleet's control of the Caspian Sea, opening a new maritime route for the Russian conquest in Central Asia. The Persian Empire's traditional influence over the eastern flank also would not last long. The intrigues unleashed by the British in Afghanistan, and the Russian advance through the Caspian Sea, would eventually cut off the Persian millenary links with this area; in 1856, the town of Herat, capital of the Persian Khorasan, a relevant oasis-town and a center for important trade routes linking the Indo basin, the Iranian plateau and the Amu Darya basin, was occupied by the troops of the Pashtun leader Dost Mohammad Khan, Kabul's governor, backed by the British. The Persian King Naser al Din Shah, angry because of the situation, orders the military occupation of the town. After a tough fight, the Persian troops conquered the site but theirs was a useless victory: the British army occupied the shore region of Busher and began the invasion of the Iranian Khuzestan (in the south-east of the country), forcing the Persian Crown to sign an agreement known as the Anglo-Persian Treaty of 1857, as a result of which Persia committed to retire its troops and to not intervene in the internal affairs of Herat.[25]

– *Persia and Russia in Central Asia.* In 1869, the Russian army set up its first naval base on the eastern shore in spite of the Persian Crown's protests. The Russians answered that they would not advance towards the south and that the base was built to control the Turkmen tribes' looting and piracy, also a problem in the eastern provinces of the Persian Empire. Once the Turkmen tribes were completely defeated in the Battle of Geok-tepe (1881), the territory roughly known today as Turkmenistan was incorporated into the Russian Empire. After the Russian victory, there was an agreement between Persia and Russia (Akhal-Khorasan Convention 1881) establishing the borders between the two empires. The limits were the Atrak river, the ZirKuh mountains and the town of Sarakhs: the same limits as the current Turkmen-Iranian border. Under pressure from both Great Britain and Russia, Iran lost most of its old peripheral provinces and, in 1907, both empires signed a secret agreement to divide Persia into two areas of influence: the north for Russia, the south for Great Britain.[26]

- *East of Amu Darya: Chinese Turkestan.* In 1866, shortly before the capture of Samarkand by the Russians, the Tajik Yakub Beg (general of the Kokand khanate) led a successful insurrection and established the independent khanate of Kashgar. In 1873, when Russia and Great Britain accepted the Amu Darya basin as the border between their empires, the Viceroy of India, Lord Northbrook, sent a delegation led by the English diplomat Douglas Forsyth to recognize the new khanate's recent independence, to establish a trading relationship and to evaluate Yakub Beg's expectations regarding Russia and China. The relations between Great Britain and the Kashgar khanate were quite hopeful, because the British expected the khanate to be an obstacle against the Chinese and Russian expansion in Central Asia, and the khan Yakub Beg relied on British support to consolidate their recent independence, which was threatened by Russia and China. The unexpected death of Yakub Beg in 1877, apparently poisoned, spoiled alliance plans, and India's new governor, Lord Lytton, considered it was more convenient to facilitate the Chinese reconquest as a counterbalance to the increasing Russian presence in Central Asia. Great Britain retained a consulate in Kashgar until the end of the First World War.[27]

In 1873, the Russians and the British recognized the Amu Darya as their common border, though its limits would not be specified until the 1895 convention because of orographic difficulties in drawing them, and the disagreements between both sides. In 1879, the construction of the Turkmenbashi-Bujara-Samarkand-Tashkent-Andijan railroad began, with a route in part parallel to the Amu Darya river, whose basin was thus strategically strengthened as a transportation corridor that allowed large armies to move faster, the speeding-up of commercial transactions (cotton exports) and the improvement of links between the main urban centers in Central Asia: railways became the essential structures of these territories and their relations with the Russian metropolis.[28]

THE WORLD WARS

During World War I (1914–1918) the strengthening of the Amu Darya basin's strategic benefits was not unnoticed by Great Britain; because of the power vacuum and anarchy existing in the Russian

Empire after the revolutionary victory of the Bolsheviks in 1917, the British Government decided to send a regiment commanded by Sir Wilfrid Malleson to take control of the above-mentioned railroad, with two purposes:

1. To prevent central powers from using the railroad to send their troops up to India; and
2. To support the white armies in their fight against the Bolsheviks.

However, the mission did not succeed and, rejected by the Red Army, they returned in August 1919 to their Mashad bases in Iran.[29]

The Amu Darya border became a controversial one again 10 years later because about half a million Turkmen, Uzbek and Tajik refugees fled from the Soviets to its southern shore, as well as some Basmachi leaders (the Basmachi put up armed resistance against the Soviet imperialism and it was a heterogeneous movement with traditional Islamists and Pan-Turkists who were angered by Soviet impositions in Central Asia. The Basmachi were organized in local guerrilla bands acting in different areas but not politically or military unified).

Some of the Basmachi leaders—Enver Pasha, Ibrahim Bek, Alim Khan, Juhayd Kahan—expected to use the north of Afghanistan as an anti-Soviet resistance sanctuary. Notwithstanding, the Afghan emirs did not support them because they feared a direct Soviet intervention. The Basmachi were eventually dissolved or eliminated in the reign of the Afghan emir Muhammad Nadir Sha.[30]

While the Basmachi fought and were defeated, the territories on the northern shore of the Amu Darya became part of the Union of Soviet Socialist Republics, a new territorial entity created under Marxist and Leninist principles and ruled by the Soviets.

The southern shore of the Amu Darya, in Afghanistan, saw the slow and difficult process of the construction of a national state initiated during the reigns of Abdul Rahman (1880–1901), Habibullah (1901–1919), Amanullah (1919–1929), Nadir Khan (1930–1933) and Mohammad Zahir Sha (1933–1973). Throughout this period the borders of the current Afghanistan were delimited by agreements between Russia and Britain, several constitutions were approved (1923, 1931, 1963) with the purpose of transforming the country into a parliamentary monarchy based on the common law, the first ministerial departments were established and the creation of a national army was undertaken. However, this national project did not become a reality, for two serious reasons:

(a) The great power and autonomy of the regional khans, legitimized by their ethnic group, their descent or their observance of tradition and Sharia; and
(b) The political and economic weakness of the central government and the continuous intervention of foreign powers in Afghanistan's internal affairs, something seen as a lack of legitimacy by its own citizens.

All these caused continuous internal tensions in the country, such as the Third Anglo-Afghan War (1929–1930) and the assassination of Nadir Sha (1933).[31] Despite this, however, the Afghan-Soviet border remained stable for almost 40 years.

THE SOVIET INVASION OF AFGHANISTAN AND THE USSR DISSOLUTION

On 27 December 1979, the Soviet Armed Forces crossed the Amu Darya and occupied Afghanistan. The aim of the invasion was to secure the Afghan regime's docility after a period of internal turbulence among supporters of the overthrown Sha Mohammad Zahir Sha (1973) and his cousin Daud. The rivalry between different communist groups and factions, together with the Islamic revolutionary triumph in Iran, provoked the Russian invasion of Afghanistan. For 10 years the Afghan territory became a battlefield where several mujahidin groups, supported by the CIA, Pakistan and Saudi Arabia, forced the Soviet withdrawal in 1989. The end of the Soviet occupation started with the outbreak of a bloody civil war between triumphant guerrilla groups and with new competition between regional actors (Iran, the USA, Pakistan, Uzbekistan and China) to give support to the different Afghan factions.[32]

In 1996, the Taliban (mainly Pashtun), supported by Pakistan and Saudi Arabia, took over Kabul, while the opposition, the so-called North Alliance (composed of different Uzbek, Turkmen and Tajik guerrilla bands, backed by Russia, Iran and China) was gradually more corralled on the Amu Darya's southern shore. The 11 September (9/11) attacks in 2001 were carried out in this context, giving way to a new occupation of Afghanistan by troops under the United Nations and NATO's mandate, once more becoming a battlefield land for NATO troops and the Afghan insurgence.[33]

At the same time, on the northern shore of the Amu Darya, the Soviet collapse gave way to the birth of five new independent states, defined by an autocratic political system, fragile economies and the outburst of numerous ethnic and civil conflicts.

GEOGRAPHY OF AMU DARYA RIVER

The Amu Darya river, with a length of 2,540 km, rises in the Pamir mountains, flows through the Turan plateau and crosses the deserts of Kyzylkum and Karakum.[34]

The Amu Darya's main water contribution comes from the mountain areas of Pamir-Alai-Hindu Kush. As a result of the rainfall pattern, and the melting of different glaciers, water flow fluctuates at between 800 mm^3 and 1,600 mm^3.[35]

The river flows along most of its course through the Karakum and Kyzylkum deserts, whose main feature is their extreme dryness with a low level of precipitation between 200 mm^3 and 75 mm^3 annually. The climate is continental, with temperatures that can drop below 0° C in winter and rise near 50° C in summer.[36] These climatic and orographic conditions thus make the water volume of both rivers decrease dramatically from east to west, as well as its distribution in the region.

The Amu Darya river begins to be known by that name after the confluence of the rivers Panj, Vakhsh and Kunduz; other important tributaries are the Surkhan and the Sherabad. The Amu Darya basin occupies an area of 1,017,800 km^2 and its annual water flow can oscillate between 70 km^3 and 80 km^3.[37]

The Amu Darya water flow is the largest in Central Asia and its course crosses more countries than any other (Tajikistan, Afghanistan, Uzbekistan and Turkmenistan), acting as the border between Tajikistan and Afghanistan, Uzbekistan and Turkmenistan, and Uzbekistan and Afghanistan.

DEMOGRAPHY IN THE AMU DARYA BASIN

The population in the Amu Darya basin in 1960 was around 14 million, and roughly 50 million in 2010. The more densely populated areas are Uzbekistan (Urgench, Qarshi, Bujara, Samarkand, Khiva and

Nukus), the south of Tajikistan (Vakhs Valley) and the north of Afghanistan (Balkh, Kuduz and Kokcha).

The most populous country in the Amu Darya basin is Uzbekistan, followed by Tajikistan, Turkmenistan and Afghanistan. The countries with large numbers living in this basin are Turkmenistan and, second, Tajikistan.[38]

There is a wide variety of ethnic groups along its more than 2,500 km, mainly Uzbeks, Tajiks and Turkmens, distributed on both shores of the river and the borders between Uzbekistan, Tajikistan and Afghanistan. There are also some minorities such as Karakalpaks in its lower basin and Kyrgyzs around the sources of its main tributaries.

WHERE DOES THE RIVER'S WATERCOURSE COME FROM AND HOW IS IT DISTRIBUTED?

As the next data table shows, 62 % of the water flowing through the river's course is generated in Tajikistan. Afghanistan's contribution is the second in volume, with 27 %, while the amounts from Uzbekistan (6.3 %), Kyrgyzstan (1.9 %) and Turkmenistan (1.9 %) are much less important. Almost 90 % of the water flowing along its shores is generated in Tajik and Afghan territories (Tables 2.1 and 2.2).

Table 2.1 Annual water flow generated by shore countries of the Amu Darya river

	Flow in millions of m³	Annual % of the total flow
Tajikistan	50,000	62.5
Afghanistan	22,000	27.5
Uzbekistan	5,000	6.3
Kyrgyzstan	1,500	1.9
Turkmenistan	1,500	1.9
Total	80,000	100

Source: The Afghan Part of Amu Darya basin. Impact of Irrigation in Northern Afghanistan on Water Use in the Amu Darya basin. Walter Klemm, Sr Land & Water Development Engineer Investment Centre Division /FAO Rome and Sayed Sharif Shobair, Chief Engineer and Coordinator of the Emergency Irrigation Rehabilitation Project of the Ministry of Energy and Water of Afghanistan, FAO Kabul. http://www.unece.org/fileadmin/DAM/SPECA/documents/ecf/2010/FAO_report_e.pdf

Table 2.2   Water distribution limits in the Amu Darya basin, following Protocol 566, 11 March 1997

	Limit (m^3 per year)	Average %
Uzbekistan	29.6	45.2
Tajikistan	9.5	15.4
Kyrgyzstan	0.4	0.6
Turkmenistan	22.0	35.8
Total	61.5	100

Source: The Afghan Part of Amu Darya basin. Impact of Irrigation in Northern Afghanistan on Water Use in the Amu Darya basin. Walter Klemm, Sr Land & Water Development Engineer Investment Centre Division /FAO Rome and Sayed Sharif Shobair, Chief Engineer and Coordinator of the Emergency Irrigation Rehabilitation Project of the Ministry of Energy and Water of Afghanistan, FAO Kabul. http://www.unece.org/fileadmin/DAM/SPECA/documents/ecf/2010/FAO_report_e.pdf

WHO CONTROLS THE AMU DARYA'S WATERCOURSE?

Despite 62 % of this watercourse being generated in Tajikistan, this country has the largest water reserve (Nurek dam: 10.5 km^2) in one of the most important Amu Darya's tributaries, the Vakhs River. This tributary provides 25 % of the Amu Darya's volume of water, but the potential control Tajikistan exercises over the Vakhs affects only about half of its water, 10.5 km^3 out of 20.5 km^3 per year, or even less, because the sedimentation process suffered by the dam has reduced its capacity to 8.5 km^3.[39]

The control over the Amu Darya's volume of water, therefore, is distributed among a huge number of dams and reserves located along its course. None of the shore countries currently has the necessary infrastructures to control its present volume of water, as happens in the case of the other great Asian river, the Syr Darya. In this watercourse, the Toktugul dam regulates around 50 % of its volume.[40]

Tajikistan and Afghanistan are the only countries of the Amu Darya basin that do not depend on their neighbors to renovate their water resources.

WATER RESOURCES MANAGEMENT IN THE AMU DARYA

The immediate origin of current tensions between the largest water consumers (Turkmenistan and Uzbekistan) and producers (Tajikistan and Kyrgyzstan) arose when both rivers stopped being managed as unitary water basins, when the Soviet republics of Central Asia won their independence and the region was broken up into five new and independent states. During the Soviet years, management was dealt with by three state agencies—the Ministry of Territorial Administration and Water Resources, the Ministry of Agriculture and the Ministry of Energy—which, together with the State Planning Commission, established the amount of water to be used and the ways and aims of distribution among the Central Asia Soviet republics (consumption quotas). Decision-making was therefore centralized in Moscow and whenever a dispute arose between the republics the arbitration was ultimately settled by Moscow.[41]

The situation changed radically when the Central Asian republics became independent. The centralized process to establish the amount of water to be consumed, its aims of use and how to distribute it among consumers broke up and the presidents of the new nations met hurriedly in February 1992 in Alma-Ata to create a new institution to regulate distribution and consumption. These efforts enabled the Agreement on Cooperation in the Field of Joint Water Resources Management and Conservation of Interstate Sources, as well as the creation of the Interstate Commission for Water Coordination of Central Asia (CWC), with the purpose of establishing consumption quotas for each country as well as the start-up of some otherbodies, such as the associations for the management of the Amu Darya and Syr Darya basins. These institutions began to work with the Fund for Saving the Aral Sea, but the above-mentioned Commission does not work with the necessary efficiency and effectiveness, especially in the case of Amu Darya.

Regarding its brother river, the Syr Darya, the cooperative effort has reached higher levels as a result of several circumstances[42]:

1. All the Syr Darya shore countries belong to the CWC, but in the case of Amu Darya, Afghanistan is not included in the institution.
2. The Syr Darya watercourse is regulated basically by one country, Kyrgyzstan; control over the Amu Darya's course is more complex.
3. In the Syr Darya, unlike in the Amu Darya, there is a point of agreement between Uzbekistan, Kazakhstan and Kyrgyzstan in the field of energy water use.

The agreements on the Amu Darya, signed in 1992, turned out to be fragile and fruitless, giving way to a situation of permanent uncertainty: each part involved in water management generally puts its own interest before the global ones, so that there are frequent conflicts and crises between the region's countries, essentially with regard to water use and allocation, as the next table shows (Table 2.3).

Table 2.3 Potential conflicts in the Aral basin and their nature

Basin desiccation	Countries sharing the basin	Potential conflicts	Nature of the conflicts
Amu Darya	Tajikistan (Higher), Uzbekistan (Medium), Turkmenistan (Lower), and Karakalpakstan in Uzbekistan (Lower)	Between two states down the river, Turkmenistan and Uzbekistan, regarding the Karakum channel management	A question of volume
		Other potential conflicts on water distributed in the lower Amu Darya between Uzbekistan y Turkmenistan	A question of quality
		Between users in Higher Tajikistan (i.e. Vakhsh Valley) and users in Lower Chardzhou and Dashhowruz, provinces in Turkmenistan and Khorazmoblast in Uzbekistan and Karakalpakstan	Questions of volume and quality. Potentially long conflicts between energy versus irrigation
Zarafshon		Between users in Higher Penzhikent, Tajikistan, and users in Lower Samarkand, Bujhoro, Uzbekistan	A question of volume

Source: Ecologic Institute for International and European Environmental Policy. Pfalzburger Str. 43/44, 10717. Berlin, Germany. Transboundary river basin management regimes: the Amu Darya basin case study. Filename NeWater Deliverable 1.3.1 Amudarya_final_draft. Authors Nicole Kranz, Antje Vorwerk, Eduard Interwies Document history First Draft completed 19 July 2005. http://citg.tudelft. nl/fileadmin/Faculteit/CiTG/Over_de_faculteit/Afdelingen/Afdeling_watermanagement/Secties/ waterhuishouding/Leerstoelen/Waterbeheer/People/old/Raadgever,_G.T./doc/D131_AmuDarya_ Final.pdf

In this context, each state puts in place different strategies based on its own national interests, without any consensus with its neighbors to deal with an increasing water demand by their societies. This fact might worsen the current water-resources-related conflicts in the Amu Darya basin and could cause new ones to develop.

NOTES

1. Jaquenod, Silvia (2005): *Derecho Ambiental. La Gobernanza De Las Aguas* (Dykinson). p. 72.
2. United Nations (UN): http://www.un.org/spanish/events/water/sinfron teras.htm. Published by the UN Department of Public Information -DPI/ 2293 G -February, 2003.
3. Giordano, Meredith A. and Aaron T. Wolf. (2003): Sharing waters: Post-Rio International Water Management. *Natural Resources Forum* 27: 163–171.
4. United Nations. Human Development Report 2006. http://hdr.undp. org/en/media/07-, p. 205.
5. Aguilera, F. (1998): *Hacia una nueva economía del agua: cuestiones fundamentales*. Departamento de Economía Aplicada. Universidad de la Laguna. Tenerife (Spain), September 1998. Ed.: Instituto Juan de Herrera. ISSN: 1578-097X.
6. Waever, 1998.
7. International Fund for Saving the Aral Sea. Water Quality. http://www.ec-ifas.org/aral_basin/iuvr/water-quality/.
8. The international agreements on water management about the Amu Darya signed in 1992 have been in the main fragile and fruitless, giving way to a situation of permanent scarcity in the distribution of water in the basin: (1) There are no agreements on the energy-related use of the Amu Darya's water resources, unlike Syr Darya's; (2) All the countries linked to the Syr Darya basin are involved in the negotiations and organizations connected to the Amu Darya's watercourse management. However, Afghanistan, a country with one of the most important water contributions to its course, does not participate in the cooperation negotiations.
9. Asia Central. Historia Universal (Siglo XXI). Vol. 16, Compiledby Gavin Hambly, 1977.
10. Ibid.
11. Ibid.
12. Ibid.
13. Geiss, Paul (2003): *Pre-Tsarist and Tsarist Central Asia: Communal Commitment and Political Order in Change*. (Routledge).
14. Ibid.

15. The creation of this border speeded up the political detachment of northern Afghanistan's territories from the Central Asian khanates. This process was finalised with the independence of the Afghan state in 1919.
16. Khodarkovsky, Michael (1992): *Where Two Worlds Met: The Russian State and the Kalmyk Nomads 1600–1771*, (Cornell University Press), p. 159.
17. Zavialov, P. (2007): *Physical Oceanography of the Dying Aral Sea*. (Springer Science & Business Media), p. 11.
18. Hambly, G. (1977): Asia Central, *Historia Universal, vol. 16* (Siglo XXI), p. 214.
19. Pierce, R. (1960): *Russian Central Asia, 1867–1917: A Study in Colonial Rule*. (University of California Press), pp. 21, 31, 116.
20. Geiss, Paul (2003): *Pre-Tsarist and Tsarist Central Asia: Communal Commitment and Political Order in Change*. (Routledge).
21. Ibid.
22. See note 16.
23. Becker, S. (2009): *Russia's Protectorates in Central Asia: Bukhara and Khiva, 1865–1924*. (Taylor & Francis), p. 107.
24. Fremont-Barnes, G. (2009): *The Anglo-Afghan Wars 1839–1919* (Essential Histories) [Paperback] (OSPREY).
25. Amanat, A. (1997): *Pivot of the Universe: Nasir al-Din Shah and the Iranian Monarchy, 1831–1896* [Hardcover]. (University California Press), pp. 259–309.
26. Siegel, Jennifer (2002): *Endgame: Britain, Russia and the Final Struggle for Central Asia*. New York. (I.B. Tauris), pp. 1–21.
27. Cyril E. Black, Louis Dupree, Elizabeth Endicott-West, Eden Naby (1991): *The Modernization of Inner Asia*. (M.E. Sharpe), pp. 31–51.
28. Sahedeo, J. (2007): *Russian colonial Society in Tashkent, 1865–1923*, (Indiana University Press), p. 190.
29. Major-General Sir Wilfrid Malleson. The British military mission to Turkistan, 1918–2. *Journal of the Royal Central Asian Society*. Volume 9, Issue 2, 24 January 1922, pp. 95–110.
30. Rywkin, M. (1929–1933): *Moscow's Muslim challenge: Soviet Central Asia*, p. 43.
31. Omrani, Bijan (2007): Afghanistan and the Search for Unity, published in *Asian Affairs*, Volume 38, Issue 2, 2007, pp. 145–157.
32. Dixon, Norm (2001): *Revolution and counter-revolution in Afghanistan*. http://www.greenleft.org.au/2001/475/24709.
33. Kaplan, Robert D. (2001): *Soldiers of God: With Islamic Warriors in Afghanistan and Pakistan*; (Vintage Departures), p. 166.
34. Nicole Kranz, Antje Vorwerk, Eduard Interwies (2005):, Ecologic Institute for international and European Environmental Policy, Pfalzburger Str. 43/44, 10717 Berlin, Germany. Transboundary river basin management regimes: the Amu Darya basin case study. Filename NeWater Deliverable

1.3.1. Amudarya_final_draft. http://citg.tudelft.nl/fileadmin/Faculteit/
CiTG/Over_de_faculteit/Afdelingen/Afdeling_watermanagement/
Secties/Waterhuishouding/Leerstoelen/Waterbeheer/People/old/
Raadgever,_G.T./doc/D131_AmuDarya_Final.pdf.

35. Loktionova, M. (1992): *Large-scale peculiarities of the global distribution of
    snow cover, Polar Geography and Geology*, 16:2, 148–159. To link to this
    article: http://dx.doi.org/10.1080/10889379209377482.

36. Abazov, Rafis (2008): *Palgrave Concise Historical Atlas of Central Asia*.
    (Palgrave Macmillan).

37. Nicole Kranz, Antje Vorwerk, Eduard Interwies (2005): Ecologic Institute
    for international and European Environmental Policy, Pfalzburger Str. 43/
    44, 10717 Berlin, Germany. Transboundary river basin management
    regimes: the Amu Darya basin case study. Filename NeWater Deliverable
    1.3.1. Amudarya_final_draft. http://citg.tudelft.nl/fileadmin/Faculteit/
    CiTG/Over_de_faculteit/Afdelingen/Afdeling_watermanagement/
    Secties/Waterhuishouding/Leerstoelen/Waterbeheer/People/old/
    Raadgever,_G.T./doc/D131_AmuDarya_Final.pdf.

38. UNEP_GRIDA Environment and Security in the Amu Darya River Basin
    2011. Available online: http://www.grida.no/publications/security/
    book/4881.aspx.

39. Interstate Water Resource Risk Management. Towards A Sustainable Future
    for the Aral Basin. Editor(s): Oliver Olsson and Melanie Bauer. Publication
    Date: 10 February 2010. ISBN: 9781843393085, p. 200. Paperback p. 76.

40. Votrin, V. (2003): Transboundary Water Disputes in Central Asia: Using
    Indicators of Water Conflict in Identifying Water Conflict Potential, p. 7.
    http://www.transboundarywaters.orst.edu/publications/related_
    research/Thesis_Votrin.pdf.

41. Sarah L. O'Hara (2000): Historical Perspectives on Global Water.
    Challenges Lessons from the past: water management in Central Asia.,
    Water Policy Volume 2, Issues 4–5, pp. 365–384. (School of Geography,
    University of Nottingham, Nottingham, NG7 2RD, UK). Available online
    16 October 2000.

42. Nicole Kranz, Antje Vorwerk, Eduard Interwies (2005): Ecologic Institute
    for international and European Environmental Policy, Pfalzburger Str. 43/
    44, 10717 Berlin, Germany. Transboundary river basin management
    regimes: the Amu Darya basin case study. Filename NeWater Deliverable
    1.3.1. Amudarya_final_draft. http://citg.tudelft.nl/fileadmin/Faculteit/
    CiTG/Over_de_faculteit/Afdelingen/Afdeling_watermanagement/
    Secties/Waterhuishouding/Leerstoelen/Waterbeheer/People/old/
    Raadgever,_G.T./doc/D131_AmuDarya_Final.pdf.

CHAPTER 3

Environmental Security, Water Resources and International Security Organizations

Abstract There is a wide range of environmental threats to the Human Right to Water and Sanitation (HRWS) in Central Asia: desertification, the intensive use of insecticides and fertilizers harmful to the environment, global warming, nuclear pollution risks—uranium waste dumps—or pollution coming from extractive or military industrial complexes. Most of these threats are of a transnational nature and are the cause or consequence of water resource management models. In this chapter, the author analyzes the conflicts, risks and vulnerabilities in the region related to the security environment and water resources, as well as the cooperative response to the environmental problems by international organizations—the Collective Security Treaty Organization (OTCS), the Shanghai Cooperation Organisation (SCO), Organization for Security and Co-operation in Europe (OSCE) and NATO-sponsored programs—in terms of security policies.

Keywords Contamination · Salinization · Climate change · Speech Act · Agent · Water securitization process

Jorge Nef (1999) defines environmental security as the right of individuals and communities to the preservation of their life and health, and to dwell in a safe and sustainable environment. One of the links between these rights of individuals and communities is water.[1] This has been interpreted

© The Author(s) 2017   39
M.Á. Pérez Martín, *Security and Human Right to Water in Central Asia*, Security, Development and Human Rights in East Asia,
DOI 10.1057/978-1-137-54005-8_3

as requiring parties to ensure an adequate supply of safe and potable water and basic sanitation, and to take steps to protect its population from exposure to harmful chemicals or other environmental contaminants.[2]

Water, environment, human security, human rights and emancipation are linked indissolubly. All human beings depend on the environment in which they live. A safe, clean, healthy and sustainable environment is integral to the full enjoyment of a wide range of human rights, including the rights to life, health, food, water and sanitation. Without a healthy environment, we are unable to fulfill our aspirations or even live at a level commensurate with minimum standards of human dignity. At the same time, protecting human rights helps to protect the environment.[3]

Poor management of the environment has an impact on the amount and quality of the available water resources, and vice versa. Bad management of water means serious environmental damage, jeopardizes human security and violates the human right of access to clean and sufficient water, thus reducing the chances of the empowerment and emancipation of individuals and communities.

In this chapter we shall analyze the main environmental threats regarding water, and the outcome of national and multilateral strategies to neutralize those threats.

THE SOVIET UNION, WATER AND ENVIRONMENT

For more than 60 years, the environment and natural resources in the Soviet Union, including water, were subject to the military-industrial complex whose priority was to compete in military and material terms with the Western bloc, trying to survive as well as to uphold the world supremacy of the USSR; in consequence, environmental matters were not relevant, and the preservation of the quantity and quality of natural resources was not of first importance. Any criticism of this situation would have been considered an attack on national security.

The inclusion of the Central Asian region in the Soviet Union produced some drastic changes ton its environment, affecting irreparably the amount and the quality of the available water resources from its two largest rivers, the Amu Darya and the Syr Darya. While the USSR had many laws and decrees to regulate the preservation of the environment,[4] these were subject to, or ignored by, the Soviet nomenklatura (Party appointees to government posts), with the aim of building large dams and canals devoted to supplying great amounts of water for cotton production and the development of a highly polluting extractive and energy industry. These were political decisions that have led to serious environmental disasters in recent times, such as the desiccation of the Aral Sea.

ENVIRONMENT, WATER AND THE ARAL SEA

The baseline for the Aral Sea can be considered as its levels before the 1960s, as it remained relatively stable from 1901 to 1961 (United Nations Environment Programme 2014). During this time, it covered approximately 66,000–67, 500 km^2 (depending on source) in area and held about 1,060 km^3 of water (United Nations Environment Programme 2014). The Amu Darya and Syr Darya rivers contributed a combined 47–55 km^3/yr of water (United Nations Environment Programme 2014). In less than a generation, however, it had shrunk to 10 % of its former volume. The diversion of water for cotton began during the Soviet era, and has continued after the collapse of the USSR. The Aral Sea's desiccation has been produced by the overexploitation of the rivers that supply it (Syr Darya and Amu Darya), making one of the worst environmental catastrophes of modern times, whose consequences have affected mainly the Kazakh province of Kyzl-Orda and, in particular, the Uzbek one of Karakalpak. Below are listed the regional keys to understanding the environmental effects:

(a) **Rising temperatures, lower rainfall**. The Aral Sea used to regulate the region's climate. The disappearance of a great part of its water volume has given way to a decreasing environmental humidity, of about 10 %, and an increasing evaporation index, up to 1,700 mm^3 per year. Winters in the region are now colder and summers are hotter, between 2°C and 3°C degrees higher, with record temperatures occurring of 49°C. This has caused a gradual lowering in the amount of rainfall, and the more immediate consequence has been a growing desertification of the environment and the disappearance of a great part of the farming, stockbreeding and fishing activities in the region.

(b) **Economy**. One of the key sectors, fishing—the region's economic driving force—has disappeared. The increasing salinity of the Aral Sea meant an environmental catastrophe that led to the death of the Kazakh and Uzbek fishing sector here. At the beginning of the 1960s the sea's salinity level was roughly 1 %, and 44,000 tons of fish were captured annually. The fishing wealth, employing 60,000 people, disappeared completely. Around 100,000 people have been forced to leave the area.[5] After a continuous water level drop in the 1980s, salinity in the Aral Sea increased from 1 % to 10 % per liter.

(c) **Health.** Three million hectares of marine soils containing phosphates and heavy metals, coming from farming and industrial wastes, have been exposed to winds and erosion after the Aral Sea's desiccation, polluting the air and the aquifers destined for human use. In Karakalpak, the Uzbek province where the Amu Darya delta is located (to the south of the Aral Sea), there are about 1.2 million people, half of them belonging to the Karakalpak ethic group. The data on Karakalpak's population are awesome: 30 % of newborns show malformations, and one in 10 children is stillborn; 120 mothers in100,000 die in childbirth; and cases of typhus and hepatitis have increased by 30 times since 1980. In addition to these data, there is an infant mortality rate of 35 in 1,000 children, and a general life expectancy of less than 38 years. Tuberculosis is active, with 167.9 cases per100,000 inhabitants; the World Health Organization (WHO) establishes that 50 cases per 100,000 are considered as an epidemic disease. Some estimates suggest that in the period 1981–1987, liver cancer rates grew by an unbelievable proportion of 200 %; with 25 % in the case of throat cancer; and there was a 20 % increase in infant mortality. Local doctors claim that the blame lies with water pollution. The water consumed by most of the people is polluted by water coming from drains, saturated by concentrated chemical substances from cotton cultivation fields, and salts. In the Kazakh province of Kyzil-Orda, with 650,000 inhabitants, the situation is a little less dramatic, but the data are also conclusive, especially in the coastal districts of Aralsk and Kazalinsk, where the average life expectancy is around 40 years.

As shown by documentary evidence, the outcome of water resources management during the Soviet era was catastrophic for the Aral Sea and the people living in its environments. The progressive desiccation of the Aral Sea produced a deep economic, environmental and health crisis without precedent in the region.

These conditions of environmental and human catastrophe are not comparable to the rest of the Amu Darya basin, but the situation regarding the quality of water and its safety is a cause of great concern. Despite the political and socioeconomic changes in the region after the dissolution of the USSR, and the inclusion of the former Soviet republics as independent states among the international community, the environmental risks continue to threaten extensive areas in Central Asia, and the degradation of the amount and quality of the resources in the Amu Darya basin is evident.

Agricultural Pollution

Decades of defective irrigation practices and the excessive use of chemical fertilizers and insecticides have seriously affected, and are still affecting, the quality of water and soils in Central Asia. Irrigation plans for extensive areas without a previous planning study of soil, water, geomorphology and hydrology being done, have been to blame for the salinity of huge areas of agricultural land in Central Asia.

The salinization process has been produced in part by an irrigation technique consisting of just flooding a specific plot of land, thus generating a high concentration of salts and significant soil erosion.

Another source of pollution is the intensive use of large amounts of pesticides and fertilizers, especially the most soluble ones, even though they are not necessary to cultivation, thus producing high concentrations of salts that contaminate the aquifers.

As the following table shows, large areas of land are affected by salinization (Table 3.1).

The countries most affected by salinization are Turkmenistan and Uzbekistan— with 95 % and 50 % of their soil, respectively—and then Kazakhstan, 33 %, Tajikistan, 16 %, and Kyrgyzstan, 11 %. The continuous

Table 3.1   Soil salinization in Central Asia

	Irrigation area (hectares)	Area affected by salinization	Irrigation area (%)
Kyrgyzstan	1,077,100	124,300	11.5
Tajikistan	719,200	115,000	16.0
Kazakhstan	2,313,000	>763,290	>33.0
Turkmenistan	1,744,100	1,672,592	95.9
Uzbekistan	4,280,600	2,140,550	50.1
Central Asia	10,134,000	4,815,732	47.5

Source: Ministers tvookhr any prirody Turkmenistana for UNEP, 2000, Doklad poo sushch est vleniiu Natsional'noi program y deist vii pobor'be s opusty nivaniem v Turkmenistane, p. 24; United Nations Environment Programme 2014, Aquastat (figures are for 1993–1994); TACIS (2000) Kyrgyz Republic National Irrigation Strategy and Action Plan. Supporting Document, pp. 2–13. Irrigation in Central Asia: waterwiki.net/.../World_Bank_Irrig_Report_-_C...

degradation by salinization of irrigated soils eases pollution by the filtration of both surface and subterranean waters, rendering them unfit for human use, degrading farming soils, which become gradually less productive, and cultivation polluted by salts. In conclusion, it puts at serious risk the productivity and sustainability of farming practices in the region. The most important international agreement regarding the fight against desertification is the Convention to Combat Desertification. The main purpose of the agreement is to promote practical action through local innovative programmes and international support cooperation; to establish guidelines to fight against desertification; or to mitigate the consequences of severe droughts, or desertification, in countries affected by them. It also emphasizes popular participation and the creation of conditions to help local populations to be self-sufficient in preventing soil degradation, assigning non-governmental organizations an unprecedented involvement in the preparation and carrying out of programs to prevent desertification. All Central Asian countries have ratified the convention to combat desertification, but its implementation has not been effective.

The only commitment reached so far by all the countries in the region regarding this matter is the creation of Cap-Net. This is an international network for capacity building in integrated water resources management. It is an UNDP program, and an associated program of the Global Water Partnership. Cap-Net's direct partners are regional- and country-level capacity building networks working in integrated water resources management. These networks are composed mainly of capacity building institutions such as universities and training institutes, but also NGOs (non-governmental organizations), government agencies, private companies, development program, etc.

The Central Asian network will help to develop local capacity to deal with a range of complex water issues, including the joint management in the Aral Sea basin of water for irrigation downstream and energy needs upstream, increasing climate change variability.

INDUSTRIAL POLLUTION

The extractive and metallurgy sector is a driver of the industrial sector of Central Asia, but it often creates vast quantities of hazardous waste such as POPs (persistent organic pollutants). POPs are among the most dangerous pollutants: they include pesticides, industrial chemicals and unwanted byproducts of industrial and other processes, are highly toxic and long-lasting, and cause an array of adverse effects, including disease and birth defects in both humans and animals. Some of the severe health impacts from POPs include cancer, damage to the central and peripheral nervous systems, reproductive disorders, and the disruption of the immune system. These impacts do not respect international borders, and are often intergenerational, affecting both adults and their children. POPs can affect people and wildlife even at very low doses. Most of the waste originated by industry and urban environments located in the Amu Darya basin are not treated adequately.

Turkmenistan and Uzbekistan have refused to sign the Stockholm Convention on Persistent Organic Pollutants, and Central Asian countries' moves to limit or ban the use of certain pesticides and fertilizers harmful for the environment have been quite poor. But Kazakhstan, Kyrgyzstan, Tajikistan and Afghanistan have signed the Convention and have developed a series of national plans with several priorities:

- To develop a national normative and legal framework to fulfill the obligations related to the Convention.
- To Include the polluting substances inventory published by the Convention in environmental monitoring systems.
- To develop specific programs to eliminate or reduce polluting products.
- To carry out viability studies to reduce pollutants in territories already contaminated.
- To carry out monitoring studies of polluting substances; devise a network of national laboratories and centers devoted to the management and treatment of polluting substances.

Despite the fact that these three countries have ratified the Convention, the implementation of their national plans is still at a very early stage. These countries have also signed the Rotterdam Convention on the Prior Informed Consent Procedure for Certain Hazardous Chemicals and Pesticides in International Trade (PIC), applicable to certain pest control substances and hazardous chemical products in international trade.

According to the Rotterdam Convention's website, Kyrgyzstan and Kazakhstan, with regard to their belonging to the Convention, have banned the import and, in consequence, the production of, chemical substances hazardous to human health and ecosystems. In Kazakhstan, the prohibition affects up to 16 chemical products, and 15 in the case of

Kyrgyzstan, though there is still many substances in the Convention list that are not regulated in either of the countries.

CLIMATE CHANGE

Climate change is a threat to the global environment. Its main effects are a rise in temperature on the earth's surface and alterations to rainfall patterns, which cause variations in the amount and quality of available water resources. From the second half of the nineteenth century to the present day, the global temperature has risen by between 0.3° and 0.6° Celsius, and these changes have also affected Central Asia.[6] This temperature rise has essentially contributed to the reduction of the area covered by glaciers in Tajikistan and Kyrgyzstan, where the region's main water reserves are generated. In Tajikistan, water coming from glaciers represents between 10 % and 20 % of the volume of the rivers flowing through its territory, but during years of drought, water from glaciers represents 70 % of that volume. Since 1930, the area of the glaciers in this country has decreased by approximately one-third.

Kyrgyzstan does not show any better data; 4 % of its territory is covered by glaciers; and the area of these is estimated to be reduced by between 30 % and 40 % if temperatures continue to rise.[7]

In general, the impact of global warming on the Aral basin means a drastic decrease in its water resources and the water volume of its rivers; there are estimates suggesting that the two main rivers of the region, Amu Darya and Syr Darya, will lose, respectively, between 10 % and 15 %, and between 6 % and 10 %, of their volume.[8]

The progressive scarcity of water resources, the increasing salinization and soil compaction have had, among other things, these consequences:

- A decreasing productivity of the land and, consequently, of the quality and size of harvests. For example, the productivity of some crops, such as cotton, has decreased by between 60 % and 50 %, barley between 30 % and 40 %, corn between 40 % and 60 %, and wheat between 50 % and 60 %. Cropland productivity has decreased, in general, in all the countries: 30 % in Uzbekistan, 40 % in Turkmenistan, 18 % in Tajikistan, 30 % in Kazakhstan and 20 % in Kyrgyzstan.
- Regarding the quality of both surface and subterranean waters, roughly 44 % of the available water supplements in the region are catalogued as

"moderately contaminated" and only 23 % as "clean or barely contaminated water". The main pollutant, though not the only one, is salts.[9]

The Convention on Climate Change is the frame in which the Kyoto Protocol was developed, with the aim of reducing the emissions of six gases that are suspected to causing global warming: carbon dioxide (CO_2), methane (CH_4), nitrous oxide (N_2O), plus three industrial fluoridated gases: hydrofluorocarbons (HFC), perfluorocarbons (PFC) and sulfur hexafluoride (SF_6). All the Central Asian countries have ratified this protocol, Kazakhstan being the last one to do so. Uzbekistan, Tajikistan and Kyrgyzstan, which have not signed any of the Protocol annexes, have a weaker commitment and are framed within the so-called "projects of clean development mechanisms", whose aims are:

(a) Improve energy efficiency in both industry and homes.
(b) Develop renewable energy sources.
(c) Use less polluting combustibles.
(d) Reforestation.
(e) Management of organic wastes from the farming sector.

These projects are carried out by the countries themselves, with the aims noted earlier, and the United Nations Framework Convention on Climate Change (UNFCCC) decide whether it finances them or not. Kazakhstan has asked to join Annex B of the Kyoto Protocol in order to reduce its CO_2 emissions, but the petition is still being examined at the time of writing.

The national implementation of the range of analyzed conventions is insufficient and slow, for several reasons:[10]

- Lack of coordination between the different agencies responsible for the implementation of the programs.
- Insufficient mechanisms to control the fulfillment of the programs and their implementation.
- Lack of human resources with regard to environmental questions.
- Lack of cross-border cooperation.
- Insufficient funding.
- Lack of empathy among Central Asian people for sustainable development concepts and practices.
- Need to adapt new environmental policies respecting, or coordinated with, traditional techniques of farming and fighting desertification.

WASTE DUMPS AND MILITARY COMPLEXES

The mining and milling of uranium ore in the Central Asian region between 1944 and 1995 provided approximately 30 % of the uranium production of the Soviet Union. This resulted in the significant legacy of about 1 billion tons of waste from this processing. After 1995, many of the conventional uranium mines were closed. There are many waste dumps (48) and uranium mines abandoned in the region without appropriate security measures, as well as being in a deplorable state of maintenance, so that the accumulated waste runs the risk to coming to the surface or filtrating into nearby streams and rivers, because of soil erosion, frequent floods in the area, or sometimes earthquakes. The most dangerous waste dumps are the worst kept ones and their locations are shown in the following tables (Tables 3.2 and 3.3).

In 2009 and 2012 there were two important meetings to discuss the storage of toxic waste, processing and transportation in Central Asia, one of them held in Bishkek, Kyrgyzstan, and the other in Genoa, Italy; several states were represented at these meetings—such as the Czech Republic, Finland, Germany, Japan, Norway, the Russian Federation, Switzerland and the USA, and some international and regional bodies and financial institutions, including the United Nations Development Programme (UNDP), the International Atomic Energy Agency (IAEA), the Organization for Security and Co-operation in Europe (OSCE), the European Union (EU), the Eurasian Economic Community(EAEC), the Global Environment Facility (GEF) and the World Bank (WB), with the aim of stimulating cooperation among the region's countries

Table 3.2 Number of waste dumps, tons of nuclear waste and country distribution

	Number	Total in each country (millions tons)	Affected area of land (km²)
Kazakhstan	3	283	51.7
Kyrgyzstan	34	86.3	6.5
Tajikistan	10	170	3.0
Uzbekistan	1	73.2	2.8
Central Asia	48	812.6	64

Source: Regional Overview Central Asia (CA) Facts about radioactive waste stores and waste dumps in Central Asia. Tailings sites. Waste dumps of rocks and low-grade ores. Country. www.un.org.kg/index2.php?option=com_resource&task=show...id

Table 3.3 Pollution risk levels of Central Asian waste dumps

Country, district	Environmental pollution risk	Risk of dumps being affected by natural phenomena	Level of transnational environmental risk	National and transnational total risk
Kyrgyzstan, Mailuu-Suu	High	High	High	High
Kyrgyzstan, Min-Kush	High	High	Medium	High
Kyrgyzstan, Kadji-Sai	Medium	Medium	Low	Medium
Kyrgyzstan, Sumsar-Shekaftar	High	Medium	High	High
Kyrgyzstan, Ak-Tyuz	High	High	High	High
Tajikistan, Khudjand Chkalovsk	High	Medium	High	High
Tajikistan, Taboshar	High	Medium	High	High
Uzbekistan, Charkesar	High	Medium	Medium	Medium

Source: Regional Overview Central Asia (CA) Facts about radioactive waste stores and waste dumps in Central Asia. Tailings sites. Waste dumps of rocks and low-grade ores. Country. www.un.org.kg/index2.php?option=com_resource&task=show ... id

(Kazakhstan, Tajikistan, Uzbekistan and Kyrgyzstan) regarding the elimination and maintenance of waste dumps.

These were the conclusions of the Bishkek meeting:

1. Central Asian countries need to strengthen their national laws on toxic wastes and harmonize them with their neighbors'.
2. Current funding resources are not enough to cope with the problem, in spite of several programs being carried out in the region at present. The World Bank has approved a sum of US$6.9 million to finance the recovery of the Mailuu-Suu dump. The Czech Republic, USA and Germany have added US$600,000 for technical support. Rosatom (a Russian company), in the context of the EAEC, has approved a total aid sum of US$28 million for the clean-up of some dumps in Kyrgyzstan and Tajikistan. The International Science and

Technology Center (ISTC) has invested US$4 million in programs related to security and the clean-up of dumps.

3. At present, there is neither an accurate inventory of the quality and quantity of toxic materials stored, nor reports about environmental and geotechnical security—and nor are there monitoring and human resources systems set up to manage them.

4. To prevent terrorists from taking over these wastes, both the security and clean-up of the dumps should be reinforced.

Considering the relevance of the problem, as well as the seriousness of the situation, UNECE organized a second meeting a few months later in Genoa, with the aim of attracting aid and establishing assistance programmes covering several areas:

– The development of institutions and the harmonization of legislation on toxic wastes in the region.
– Surveys related to radioactive pollution risks in a cross-border environment.
– Improvement in the Management of radioactive dumps.
– Dealing with the socioeconomic problems of territories, and people living near dumps and abandoned uranium mines.
– Stimulate private sector participation in the management of the radioactive legacy in Central Asia.

These dumps are undoubtedly the main transnational threat to economic and environmental security in the region, because of their capacity to contaminate watercourses. However, interregional cooperation and financing by the international community are still not well enough developed after two decades of discussions.

In 2013, and with the aim of intensifying international cooperation and reminding the world of the great environmental threat still posed by these deposits against the human security of Central Asian communities, the United Nations published Resolution A/C.2/68/L.36, stating "considering that, despite the efforts made by the States of Central Asia at the national level and despite the support of international programmes and projects to remediate former uranium mines and tailings ponds, a number of States continue to have serious social, economic and environmental problems associated with former uranium mines and tailings ponds".

In June 2015, a new fund set up by the European Bank for Reconstruction and Development (EBRD) at the request of the European Commission (EC) will finance projects to rehabilitate former uranium mines and processing sites in Central Asia. The EC provided the Environmental Remediation Account for Central Asia with an initial €8 million (US$9 million) in funding, and with further amounts under consideration. The money will be used to finance the remediation of sites in the Kyrgyzstan, Tajikistan and Uzbekistan that have already been identified as "high priority".[11]

Another issue hovering over the dumps problem is the existence of new mine projects in Kazakhstan, Uzbekistan and Kyrgyzstan, and it is not clear whether these new mining sites will have proper security measures or will just be a new pollution legacy for the region. There are some worrying cases, such as the Kumtor mine in Kyrgyzstan. The EBRD has a long-standing history with this mine, as it provided part of the initial investment for the project's development. In 2010, the EBRD signed a revolving debt facility with Centerra Gold for US$150 million, of which US$76 million has been disbursed, as stated in Centerra's most recent report for this year. As part of the deal, the bank has expressed its commitment to "high standards of transparency, environmental, health and safety conduct" and to "support the development of the Extractive Industries Transparency Initiative in the Kyrgyz Republic". These stipulations are in line with the EBRD's 2008 environmental and social policy, which emphasizes strongly "compliance with EU environmental standards", and the promotion of "good practices among the Bank's clients".

However, a recent research report on Kumtor by the *Asia Times* has revealed severe irregularities regarding the environmental impact of gold mining on the Davydov and Lysyi glaciers that also feed into Central Asia's trans-boundary water resources, the Naryn and Syrdarya rivers. Preliminary findings indicated that Centerra has been operating the mine, evidently until 2012, without a permit for waste disposal from the Kyrgyz authorities. The company's record in the Kyrgyz Republic poses the question of whether the EBRD ignored its own environmental standards regarding the Kumtor project.[12]

THE MILITARY COMPLEXES

The main environmental danger of these complexes is their pollution risk regarding river basins, subterranean waters and farming soils, because of the heavy metals and chemical substances coming from different military complexes, some of them already closed down, as in the case of the Semipalatinsk center.

The Semipalatinsk complex is located in north-western Kazakhstan, quite close to the borders of Mongolia and China; it was devoted to the study of and experimentation with nuclear bombs during the Soviet era: 456 atomic bombs were detonated, some above ground, others underground. The center was closed in 1991 by the Kazakh president Nazarbaev. Because of these detonations, according to a 2008 survey carried out by Japanese and Kazakh doctors, 220,000 people near this site received radiation levels many times above those established as compatible with human health, and pollution is still present in the soil and water. As a consequence, the region's cancer levels are between 25 % and 30 % higher than in the rest of the country, and the percentage of children born with all kinds of impairments is also above the country's average.

This complex was closed because of its highly harmful effects on human health, but there are some others still operating, such as Kapustin Yar, Lira, Baikonur Cosmodrome, Taysogan, Azgir, etc., related to experimentation on weapons of mass destruction (underground nuclear explosions), aerospace studies and missile launches, but we have not found any information about the danger of these activities for the environment, health or natural resources, possibly meaning that these complexes are continuing to pollute large areas of Kazakhstan.

We shall now analyze the role of international security organizations regarding the environment.

INTERNATIONAL SECURITY ORGANIZATIONS AND THE ENVIRONMENT

The dissolution of the USSR raised new hopes about the possibility that the new countries would apply new national policies in order to obtain the sustainable management of natural resources, less harm to the environment, an increase in transboundary cooperation, and enable the reparation of great ecological disasters produced in the Soviet Era, such as the Aral Sea's desiccation. The existence of new independent states in Central Asia led to the breakdown of Soviet strategies and concepts in terms of security, and to a new vision about the main threats in the region. In the 1990s a range of international security organizations gradually focused on this new political space (NATO, OSCE) and new ones were born (the Shanghai Cooperation

Organisation—SCO; the Commonwealth of Independent States—CIS; and the Collective Security Treaty Organization CSTO) with the aim of offering their collaboration to the new countries of the area. Their purpose was to help in preventing and solving the region's new conflicts.

The Commonwealth of Independent States (CIS)

The CIS was the organic structure that tried to replace the failed USSR, or at least that was what its founders proclaimed. In theory, this organization had the aim of managing relations between the new independent countries formerly belonging to the USSR—for example, in terms of environmental issues.

The Minsk Summit

The creation of the CIS has its origins in the Minsk Summit of 30 December 1991. This date indicates the dissolution of the USSR, the birth of the CIS (Commonwealth of Independent States) and the beginning of a process of desecuritization and resecuritization of the security relations within the ex-Soviet bloc with a strong emphasis on environmental questions. The Soviet republics that signed the agreement, originally Russia, Ukraine and Belarus, proclaimed themselves to be new subjects of international law, with their own sovereignty, the right of self-determination, stating their concern about human rights, the build-up of democratic systems, and the rule of law. The new framework of relations between these new states would depend on the United Nations Charter, the Helsinki Final Act and the Conference on Security and Cooperation in Europe (CSCE), as well as on the creation of a new Commonwealth of Independent States (CIS). It opened up the possibility that other Soviet republics, or even states not from the former USSR, might join the new association of states (article 13 of the Minsk agreement).

The Minsk Summit had a strong perlocutionary effect on Central Asia, because five of its presidents met urgently in Ashgabat, Turkmenistan, on 13 December 1991, to make their protests against not being invited to the Summit and their desire to participate in the creation of the CIS as full members and with the same conditions as the founders. However, the Soviet republics' urgency to join this process and the Russian one to convene the Minsk summit had different meanings, audiences, contexts and interests.

THE CIS, AGENTS, ACTS AND CONTEXTS

The aim of the CIS was to keep the old links that existed between the ex-Soviet republics, but formulated in a different way. The Minsk agreement appeared as a constitutive framework that created and defined new actors and rules of behavior. According to the CIS's statutes, the relations between its new members would no longer be based on the Soviet peoples' brotherhood, the planned economy and the government of the Soviets, but on democratic values, the rule of law, the free market and a new status of the former Soviet republics as sovereign and independent full member states within the framework of the international community. The CIS's strategy to reach this aim was a multidimensional one, with three different branches: political, economic and military, in order to integrate the ex-Soviet republics through the creation of common institutions to regulate the political, economic or military relations of the countries formerly within the USSR area freely and by consensus among them.

The main initial feature of this organization was that its agreements were only binding for those members that signed them, not for the rest, so that their level of integration was variable and the fulfillment of its directives quite diffuse.

THE SOVIET LEADERSHIP

The unarguable leader of this desecuritization movement was the Russian Federation, the only republic with a true social capital to put an end to the Soviet Union. The acknowledgment of its leadership by the rest of the CIS members was obvious when the Council of Heads proclaimed the Russian Federation to be the legitimate heir to the international legality of the USSR. The Central Asian republics had been, and still were by then, highly dependent on the Russian Federation. Their participation in the CIS was meant to underline their independence from Russia and give them a chance to negotiate their "new economic, political and social relations," not as complete equals, but at least in a normalized way (normal politics) with all the members of the CIS and in particular with the Russian Federation. The desecuritization of relations with the Central Asian republics and the creation of the CIS meant for the Russian Federation a tactical and political movement that offered new opportunities and challenges:

1. To put an end to the old Soviet legality, its ideas, values and institutions, from which a new coup d'état might arise, as happened in August 1991.
2. To retrieve the political initiative before a large number of secessionist movements come up in the nineties in order to create a new framework, of relations with the former Soviet republics, according to the status of a great power rather than a superpower.
3. To remove the solidarity mechanisms established by the Soviet system with regard to the less favoured republics, among which the Central Asian republics were net recipients.

The Minsk process was an ambiguous and complex one from the start; most of the ex-Soviet republics saw it as a forum to discuss peacefully their split from the Russian Federation, but at the same time new relations were forged between some of its members, especially Russia, Belarus, Kazakhstan, Tajikistan and Kyrgyzstan, leading to the creation of permanent institutional structures, such as the implementation of cooperation initiatives, as in the case of CSTO, or the creation of a free-commerce area. In this process of the desecuritization of the relations between Russia and the rest of the CIS members, environmental cooperation began to be a marginal question for the institutions and agendas of the CIS. In spite of the intense debates produced around environmental issues as a result of Glasnost among Soviet public opinion in the 1980s before the fall of the USSR, as well as the mobilization of a large number of ecological groups, the strength of the environmental debate gradually declined in Central Asia.

THE CIS: FROM DESECURITIZATION TO THE DISSOLUTION OF THE ENVIRONMENTAL QUESTION

After the constitution of the CIS, environmental issues were gradually moved away from, or they lost importance within its institutions and agenda. Environmental questions, in the cooperative design of the CIS, would be managed in a specialized body called the Interstate Ecological Council. There are more than 50 specialized bodies in the CIS, each with the aim of advising the main body. Professional or scientific associations participate in the specialized bodies regarding specific questions, but they cannot impose or decide on the CIS agenda. These agencies were created after the CIS and because of agreements made by only some of its members, and not all.[13] Such a low degree of institutionalization in terms of environmental risks and threats pushed these issues into a peripheral role in in the CIS's discussions and had the consequence that the agreements and protocols signed in the context of the CIS (up to 22 agreements and three protocols on environmental issues) were not finally implemented.[14]

What Circumstances Could Contribute to the Weak Institutionalization of the Environmental Cooperation Within the CIS?

The dissolution of the USSR and the creation of the CIS put an end to the possible responsibilities or duties of the former Soviet government with regard to the many environmental disasters that had occurred in the USSR's territories. Meanwhile, the new independent Central Asian republics have acquired de facto sovereign rights over their natural resources and jurisdiction over them, and environmental protection has resulted in considerable variations in national approaches in terms of participation in international environment regimes and adherence to the treaty obligations of the former USSR.

An additional circumstance is that, according to the CIS statutes, the regulations and decisions from the main body of the CIS in any sphere are not mandatory for the member states. In practice, they are fulfilled from time to time when there are other opportune circumstances compelling the member nations to adhere to the CIS's decisions. As a result, the CIS documents have mainly become declarations and propaganda for domestic electoral purposes (Kazanstsev 2008).

In early 2012, the CIS boasted it had 1,741 signed multilateral documents—agreements, decisions, declarations and regulations—while the average rate of their implementation at the national level stood at 55–56 % (Molchanov 2012).

In the meantime, the difficult economic situation of most of the ex-Soviet republics, including the Russian Federation, after their independence caused many difficulties for the funding required by the regional cooperation in terms of environment.

THE SHANGHAI COOPERATION ORGANIZATION (SCO)

The cooperative security organization (CSO) has its origins at the end of the 1980s when Jiang Zemin's China and Michael Gorbachev's USSR began a period of military détente similar to the one that had taken place in Europe. In his speech to the 27th Party Congress, President Gorbachev maintained that "the significance of the Asian and Pacific Ocean direction is growing," and stressed the need to reduce the danger of military confrontation and "to stabilize the situation there". During the following months, the Soviets put forward a number of concrete initiatives, including proposals for a Helsinki-style regional conference on security in Asia, support for a Pacific nuclear-free zone, overtures to China for the mutual reduction of forces along the Sino-Soviet border, and pledges to withdraw Soviet troops from Afghanistan and Mongolia.[15]

In 1990, China and Russia signed an Agreement on the Guidelines of Mutual Reductions of Forces and Confidence Building in the Military Field in the Area of the Soviet-Chinese Border. The two sides pledged to "reduce their military forces to the lowest level suited to normal good neighborly relations between the two countries on an equal basis for mutual security".

This process of détente was interrupted briefly by the USSR's demise, but China and the Russian Federation, the natural successor to the USSR, immediately restarted their conversations and the new independent Central Asian republics joined them because Russia and China shared new borders with these countries.

China was one of the first countries in the world to establish diplomatic relations with the five Central Asian republics. On 27 December 1991, the government of China recognized the independence of Kazakhstan, Kyrgyzstan, Tajikistan, Turkmenistan and Uzbekistan. In early January 1992, a Chinese delegation visited Central Asia and signed a communiqué establishing diplomatic relations with the governments of the five Central Asian republics. Between 1992 and 1994, all the Central Asian presidents had visited China. In 1996, the president of China, Jiang Zemin, visited Uzbekistan, Kyrgyzstan and Kazakhstan; and in June 2000, Turkmenistan and Tajikistan. Between 1996 and 1999, leaders of China, Russia and Kazakhstan, Kyrgyzstan and Tajikistan, in Central Asia, had four meetings, held respectively in Shanghai, Moscow, Alma-Ata and Bishkek. On 25 April 1996, during Russian President

Yeltsin's visit to China, he met with Chinese President Jiang Zemin, President Nazarbayev of Kazakhstan, President Akayev of Kyrgyzstan and President Rakhmonov of Tajikistant in Shanghai and signed the Agreement on Enhancing Trust in the Military Fields among the People's Republic of China, the Russian Federation, the Republic of Kazakhstan, the Kyrgyz Republic and the Republic of Tajikistan. On 24 April 1997 during President Jiang's visit to Russia, the heads of state of the five countries met in Moscow and signed the Agreement on Mutual Reduction of Military Forces along the Border Areas. The following are the most important aspects of these agreements:

- Military forces deployed in the border region will not attack each other;
- Military exercises will not be targeted at each other;
- The scale, scope, and number of military exercises will be restricted;
- There will be information exchanges and notification of important military activities to be carried out within 100 km of the border, with individual concerns receiving proper explanations;
- Invitations will be extended to each other to observe military exercises at certain levels;
- Efforts will be made to prevent dangerous military activities; and
- Friendly exchanges between military forces and frontier guards in the border region will be strengthened.

The aim of the above-mentioned visits and agreements was to establish a new framework of relations between China, Russia and the new independent countries in Central Asia, so putting an end to the period of tension and confrontation China had had with Moscow on China's western frontier during the Soviet era.

In this context of détente and desecuritization in the relations between China, Russia and the new independent Central Asian republics is when the Shanghai Cooperation Organization arose.

SPEECH AND ENVIRONMENT

According to the SCO Charter (signed on 7 June 2002) the main goals of the Organization are: "strengthening mutual trust among member countries; promoting effective cooperation between member countries in political, economic, scientific-technical, and educational spheres, and in energy, communications, environmental protection, and other areas".

In short, this Organization set out a multidimensional concept of security built around several fields of cooperation as listed above. More specifically, the SCO, in its third article, establishes:

> To promote cooperation in water resources management and to implement joint environmental projects and mutual assistance in case of natural catastrophes or those caused by human actions."

However, at the time of writing there has not been within the organization any action related to water conflicts in the Aral basin. There have been several summits of heads of state where the matter has been treated as a conflict generator among countries, thus being obliterated in order to progress in other questions.

The SCO's final statements regarding water resources have always reflected the Central Asian countries' disposition to collaborate, but with no specific measures within the sphere of the organization, though water is constantly in the background of the SCO summits of heads of state; in fact, it makes the cooperative development of the organization more difficult. According to some sources, such as the twenty-first-century *Business Herald*, a daily paper published in Beijing, a meeting held by the organization in Tashkent had to be postponed because of strong disagreements between Uzbekistan and Tajikistan over disputes related to water resources. The SCO takes water resources cooperation into account, but it has considered it to be a conflictive issue and has decided focusing its attention on different questions that SCO agrees to expand.[16]

The lack of cooperation within the CSO in matters of transboundary water resources is not a problem exclusive to Uzbekistan and Tajikistan. China and Kazakhstan have signed some agreements related to such resources in terms of information exchange and the implementation of joint studies.[17] But many questions are still to be resolved between both countries.

According to Assem Mustafina's ideas in 'Trans boundary water issues between Kazakhstan and China' (see notes 17 and 18). Though Kazakhstan achieved some progress in addressing transboundary water issues with China, n interstate agreement for water allocation along transboundary rivers has not been reached.... However, two countries still have unresolved cross-border water issues. The Irtysh and Ili water resources are not included for negotiation.[18]

Therefore, talks on water resources remain sidelined, and negotiations on water are usually carried out within the frame of "normal politics", which show a narrower and more mundane competition between countries and even within each of them.

THE OSCE: A LARGE SECURITIZATION PROCESS

The first documents related to the need for international cooperation on environmental issues come from the so-called Helsinki process, which began in the 1970s and led to a security conference in Europe, as well as to the creation of the OSCE. The Helsinki Final Act (August 1975) states in its section on environment that international cooperation is the solution to the ecological problems that affect people's wellbeing and the economic development of their countries; it calls on different groups and social forces to assume their responsibilities in order to protect and preserve the environment.

This document also warns that experience shows that economic development and technological progress must be compatible with preservation of the environment, and historical and cultural values. The best way to reach this goal is by the prevention and preservation of the ecological balance when managing natural resources. In 1990, the CSCE published the Paris Charter, advising countries to comply with the Conclusions and Recommendations of the Meeting on the Protection of the Environment of the Conference on Security and Co-operation in Europe (Sofia 1989/Vienna) regarding the prevention and control of the transboundary effects of industrial accidents, the management of hazardous chemicals, and pollution of transboundary watercourses and international lakes.

This document explained clearly the need that "The participating States reaffirm their respect for the right of individuals, groups and organizations concerned with environmental issues to express freely their views, to associate with others, to peacefully assemble, as well as to obtain, publish and distribute information on these issues without legal and administrative impediments inconsistent with the CSCE provisions."

The Seventh Economic Forum took place in 1999, with its debates focused on "Security Aspects in the Field of the Environment". Since then, the organization's declarations on water resources have begun to focus more intensely on security, and the forum came to the conclusion "that good management of scarce freshwater resources is of utmost importance to security in the OSCE area". One of the key moments in the securitization of pronouncements within the organization was in the Porto ministerial council meeting of 2002 when it accepted the decision (Mc(10) DC/5) on the implementation of the Johannesburg Programme action. With this decision, the organization invited the United Nations Economic Commission for Europe (UNECE) and other partner organizations and specialized agencies to cooperate in elaborating the new strategy to implement the principles agreed at the Johannesburg summit.

The Johannesburg implementation plan includes a chapter especially focused on water resources. Its title is "Protecting and managing the natural resource base of economic and social development", and it underlines, among many other recommendations about the use and management of water resources, the need to carry out an action plan with technical and financial aid to reduce the proportion of people who were without access to drinking water and sanitation in 2015; this plan would be included in the Millennium development goals. Regarding Central Asia, this document established the need for setting an environmental strategy for the 12 countries of Eastern Europe, the Caucasus and Central Asia. With this aim, in December 2003 the OSCE established a partnership with the United Nations (UN) to create the Environment and Security Initiative (ENVSEC), a central tool to assess and address environmental problems that could threaten security, stability and peace in South Eastern and Eastern Europe, the South Caucasus and Central Asia.

Later, in 2007, the 15th Forum concentrated on the key challenges to environmental security in the OSCE region, among which water management was an important issue. As a result of the Forum, the Madrid Ministerial Declaration on Environment and Security was adopted, along with a Ministerial Decision on water management which recommended enhancement of the close cooperation with the UNECE and other international organizations in the sphere of water management, called upon participating states to strengthen cooperation over water management, and encouraged the implementation of international instruments and OSCE commitments. Finally, the 16th

Forum in 2008 touched on maritime and inland waterways cooperation in order to increase security and promote the protection of the environment. In 2014, water was addressed within the Forum's focus on responding to environmental challenges, and sustainable water management was selected as a priority theme of the Swiss and Serbian OSCE chairs for 2014 and 2015.

Therefore, in recent years, water resources management has begun to become a priority issue in the OSCE's environmental agenda in Central Asia. What are the contexts and agents that have enabled this process? We shall analyze those that are the most relevant.

SECURITIZATION OF WATER, SECURITIZING AGENTS AND THE DISPOSITIF; THE PLATFORM FOR CO-OPERATIVE SECURITY AND 2007 STRATEGY FOR A NEW PARTNERSHIP WITH CENTRAL ASIA, EUWI

The main securitizing agent of the OSCE's water policies in Central Asia has been the Platform for Co-operative Security. The idea behind this Platform was an EU initiative after a Russian petition in 1993. Its aim was to highlight the CSCE's effectiveness in matters of cooperation with other international organizations.

The EU proposal on an OSCE Platform for Co-operative Security outlined a set of principles d to be adhered to by all international organizations in order to "work cooperatively" with the OSCE. Almost all of these principles were later incorporated into the 1999 Platform, which was designed to serve four key functions.[19] First, it identified a set of principles to which members of other organizations and institutions should adhere "individually and collectively" to enable the OSCE to "work co-operatively" with them. Second, the Platform outlined general modalities of cooperation. Those listed were regular contacts, including meetings; a continuous framework for dialogue; increased transparency and practical cooperation, including the identification of liaison officers or points of contact; cross-representation at appropriate meetings; and other contacts. Third, "recognizing the key integrating role that the OSCE can play", the participating states offered the OSCE, as appropriate, "as a flexible frame-work for co-operation of the various mutually reinforcing efforts". Fourth, it recognized that "subregional co-operation had become an important element in enhancing security across the OSCE area" and that subregional groupings "contribute to improved security not just in the subregion in question but throughout the OSCE area".

In this respect, the OSCE was expected to "facilitate the exchange of information and experience between subregional groups and may, if so requested, receive and keep their mutual accords and agreements".

This Platform is led by other international organizations and bodies, such as the UN, EU 28 and the COE (Council of Europe), because at the OSCE Ministerial Council Meetings, only the UN, EU 28, NATO and the COE are invited to attend and make contributions (i.e. are given the floor to address the Ministerial Council). All other organizations are invited to attend the meeting and, if they wish, to make written contributions. Consultations and information exchange take place at the field level with European Commission (EC) delegations and EU Special Representatives, at the headquarters level with staff meetings and visits, and at the political level with Ambassadorial and Ministerial EU—OSCE Troika meetings. These have been the priority channels through which the EU has drawn up its environmental policy in Central Asia, helping its securitization within the OSCE's agenda. In this context, some EU policy documents have been made public, such as the 2007 Strategy for a New Partnership with Central Asia, containing specific references to cooperation with the OSCE. The strategy proposes a coordinated approach by the EU to water management (Council of the European Union 2007). It lists "environmental sustainability and water" as one of the central issues in its cooperation with Central Asia. Under this headline, the EU is promoting Integrated Water Management, the achievement of the Millennium Development Goals (MDGs) (clean drinking water and good sanitation facilities) and regional security, stability and economic development as major goals (Council of the European Union 2007, pp. 21–22). In 2015, after the MDG's validity period had expired, the EU adopted a new program, entitled Sustainable Development Goals (SDGs). The second issue is the link made by the EU Strategy on Central Asia between water and energy security. In the words of the strategy "besides oil, gas and electricity, water management is a decisive aspect of [EU] energy cooperation with Central Asia" (Council of the European Union 2007, p. 18). It is therefore in the EU's interest to promote an integrated approach to energy security in the region, where water management will be a central issue.

These platforms have had a great influence on the decisions about who organizes the OSCE's environmental cooperation with other international organizations and geographic subregional frameworks, and how it is organized, establishing values, rules and practices to be shared in terms of cooperation. In this sense, the OSCE suggests as a solution to water-related issues the promotion and support of relevant international conventions, in particular those designed to prevent and combat marine and fresh water pollution, recommending states to ratify Conventions that have already been signed, as well as considering possibilities of accepting other appropriate Conventions to which they are not parties at present. Among these Conventions, the OSCE especially recommended those sponsored by UNECE: the Convention on the Protection and Use of Transboundary Watercourses and International Lakes (1992); the Aarhaus Convention (1998); the Industrial Accident Convention (1992); the Geneva Long-range Transboundary Air Pollution Convention (1979); and the Espoo Convention (1991).

Since 2002, the OSCE has been supporting the establishment of Aarhus Centres and Public Environmental Information Centres in the OSCE area through its Office of the Coordinator of OSCE Economic and Environmental Activities (OCEEA) and its field operations. It has done so in close cooperation with the UNECE Aarhus Convention Secretariat, and in most cases with the support of the Environment and Security (ENVSEC) Initiative, an inter-agency partnership (OSCE, UNDP, UNEP, UNECE, REC, and NATO as an associate partner) dedicated to reducing environmental and security risks. As of March 2012, there were 37 OSCE-supported Aarhus Centres in 10 countries in the OSCE region. There are six centres in Kazakhstan, four in Tajikistan and one in Kyrgyzstan. The centres provide a meeting place and a link between the government, individuals, businesses, academia, judiciary and civil society. The main objective is to improve the outreach of activities and environmental information to citizens, thereby increasing their awareness and possibility of influencing their natural environment.

There are some other contexts for cooperation proposed by the OSCE, such as the ENVSEC Initiative and the EU Strategy for a New Partnership with Central Asia.

ENVSEC Initiative

This initiative, comprising UNDP, UNEP, UNECE, OSCE, REC, and NATO as an associated partner, works to assess and address environmental problems which threaten or are perceived to threaten security, societal stability and peace, human health and/or sustainable livelihoods, within and across national borders in conflict prone regions. The Initiative collaborates closely with governments, particularly foreign, defense and environment ministries, national experts and non-governmental organizations (NGOs). Together with the stakeholders, ENVSEC has carried out assessments and published reports illustrated by maps, to understand the links between environment and security in the political and socioeconomic reality of South Eastern Europe, the Southern Caucasus and Central Asia. These programmes have an essentially technical and informative component regarding environmental questions, but they are not an element of negotiation to decide transboundary water resource policies.

EU Strategy for a New Partnership with Central Asia

Another context of the European water policies with a contribution to the securitization of that resource in Central Asia was the 2009 meeting in Rome. The environmental representatives of the European Union and Central Asian countries met with the intention of bolstering the implementation of the EU Strategy for a New Partnership with Central Asia, which provides the framework for closer cooperation in the fields of environmental resources and water, driven by a shared commitment to developg and organize a long-term partnership on the basis of common objectives and undertakings to strengthen environmental sustainability in Central Asia (CA), and determined to share their experience and expertise with the purpose of supporting the Republics of Central Asia to develop an efficient and integrated management of these resources.

One of the aims of this new strategy was to constitute the EU—CA platform for enhanced cooperation: environmental governance, climate change, and sustainable water management. Regarding water, the goal was cooperation on water issues within the framework of the EUWI (European Union Water Initiative) and "Berlin Water Process". The results of this process were the EU Water Initiative National Policy Dialogues (NPDs), the main operational instrument under the European Union Water Initiative (EUWI) in the countries of Eastern Europe, the

Caucasus and Central Asia. The objective of each NPD is to facilitate the reform of water policies in a particular country. NPDs support water sector reform mainly by assisting governments in:

- Developing and implementing water strategies and legislation based on IWRM (Integrated Water Resources Management) principles;
- Strengthening intersectoral cooperation to improve water and health, and implement the UNECE/WHO (United Nations Economic Commission for Europe/World Health Organization) Protocol on Water and Health, in particular by setting and implementing targets for the whole water cycle and on the water and health nexus; and
- Developing national policies for the management of transboundary waters in accordance with the Water Convention and other international environmental instruments.

Eastern European countries, as well as Central Asian ones, began to include IWRM, or at least to debate its introduction, in their national legislation and policies. In 2008, an NPD on IWRM began in Kyrgyzstan. In 2010 and 2011, policy dialogues on IWRM were initiated in Azerbaijan, Georgia, Tajikistan and Turkmenistan. In 2013, an NPD on IWRM was launched in Kazakhstan. Thus NPDs on IWRM are at present have been implemented by the UNECE in nine countries.

Policy dialogues are based on consultations with relevant ministries, agencies and institutions (including science and academia), NGOs, parliamentary bodies, and other national and international organizations. The dialogue process is usually conducted under the leadership of a high-level government representative, such as the minister/deputy minister of the environment or the chair of the state water committee. In some countries, national Steering Committees have been established to guide and steer the NPD process. They include representatives of relevant ministries, agencies and institutions, as well as NGOs.

The Steering Committees meet at least on an annual basis at the national level. They discuss key national water policy issues and adopt decisions on NPD activities. International and donor organizations, such as the EU, the United Nations Development Programme (UNDP), the Organization for Security and Co-operation in Europe (OSCE), the World Health Organization (WHO), the World Bank (WB), the European Bank for Reconstruction and Development (EBRD), the Asian Development Bank

(ADB) and donor countries (mainly Finland, Germany, Norway and Switzerland) are invited to the policy dialogue meetings.

We have so far described and analyzed the different contexts and actors with a contribution to the securitization process of water resources in Central Asia. Now it is time to evaluate whether that securitization has been successful.

SUCCESS OR FAILURE OF THE OSCE's WATER SECURITIZATION PROCESS?

The OSCE does not have its own discourse on water resources cooperation, and there is also a lack of consensus within the organization about how to manage water resources. The OSCE's efforts have basically been focused on sponsoring international conventions and national programmes devoted to improving water resource governance, most of them sponsored by EU countries. Meanwhile, the implementation of these tools by Central Asian countries, and in particular in the Amu Darya basin, has not been very successful; in the next table we can see the ratification status of UNECE environmental instruments (conventions and protocols) in Central Asia (Table 3.4).

Kazakhstan has ratified all of the 10 above-mentioned conventions and protocols, except three; Kyrgyzstan, has ratified three; Turkmenistan, two; Tajikistan, two; and Uzbekistan, one.

One of the most important conventions on transboundary cooperation relating to questions of water resources is the Convention on the Protection and Use of Transboundary Watercourses; generally speaking, its aim is that the states will take on the following commitments:

(a) To prevent, control and reduce pollution of waters causing or likely to cause a transboundary impact;

(b) To ensure that transboundary waters are used with the aim of ecologically sound and rational water management, conservation of water resources and environmental protection;

(c) To ensure that transboundary waters are used in a reasonable and equitable way, taking into particular account their transboundary character, in the case of activities which cause or are likely to cause transboundary impact; and

(d) To ensure the conservation and, where necessary, the restoration of ecosystems.

Table 3.4 Status of ratification of UNECE environmental instruments in Central Asia

Title of the instrument	Kazakhstan	Kyrgyzstan	Tajikistan	Turkmenistan	Uzbekistan
Convention on Long-range Transboundary Air Pollution (LRTAP Convention, 1979)	11.01.2001 accession	25.05.2000 accession	—	—	—
Eight protocols3 to LRTAP Convention	—	—	—	—	—
Convention on Environmental Impact Assessment in a Transboundary Context (Espoo Convention, 1991)	11.01.2001 accession	01.05.2001 accession	-	—	—
Protocol on Strategic Environmental Assessment to the Espoo Convention (2003)	—	—	—	—	—
Transboundary Watercourses and International Lakes (Water Convention, 1992)	11.01.2001 accession	—	—	—	04.09.2007 accession
Protocol on Water and Health to the Water Convention (1999)	—	—	—	—	—
Convention on the Transboundary Effects of Industrial Accidents (Industrial Accidents Convention, 1992)	11.01.2001 accession	—	—	—	—
Protocol on Civil Liability and Compensation for Damage Caused by the Transboundary Effects of Industrial Accidents on Transboundary Waters to the Water and Industrial Accidents Conventions (2003)	—	—	—	—	—
Convention on Access to Information, Public Participation in Decision-making and Access to Justice in Environmental Matters (Aarhus Convention, 1998)	11.01.2001 ratification	01.05.2001 accession	17.07.2001 accession	25.06.1999 accession	-
Protocol on Pollutant Release and Transfer Registers to the Aarhus Convention (2003)	—	—	21.05.2003 signature	—	—

Source: Economic Commission for Europe. Strengthening Water Management and Transboundary Water Cooperation in Central Asia: the Role of UNECE Environmental Conventions. http://www.unece.org/fileadmin/DAM/env/water/publications/documents/Water_Management_En.pdf

In the Amu Darya basin, only Uzbekistan (1997) and Turkmenistan (2012) have ratified this convention. However, none of them have signed some of the important protocols that are part of these convention, some deeply related to the human right to water and sanitation, such as the Protocol on Water and Health (1992). The objective of this Protocol is to promote at all appropriate levels, nationally as well as in transboundary and international contexts, the protection of human health and well-being, both individual and collective, within a framework of sustainable development, through improving water management, including the protection of water ecosystems, and through preventing, controlling and reducing water-related disease. The Protocol is therefore very relevant for all Central Asian countries, which pay a very high social cost because of lack of access to safe water and sanitation and water-related diseases—among the most common causes of child mortality.

They have also not signed the Protocol on Civil Liability and Compensation for Damage Caused by the Transboundary Effects of Industrial Accidents on Transboundary Waters.

On analyzing the previous table, we could state that the Amu Darya basin countries choose different tools of multilateral transboundary cooperation, that they take on only limited commitments, and that some of them remain completely on the sidelines, as Afghanistan, for example, standing in the way of transboundary cooperation and implementation of multilateral agreements in matters of water resource management.

Regarding the national plans sponsored by the EUWI in the Amu Darya basin, the next table shows the different contexts of water resource cooperation with Central Asian countries that have been started, but their implementation has not yet been put in place; meanwhile, Uzbekistan, the most populous country in the region, and the one consuming the most water, is still outside these instruments of cooperation (Table 3.5).

Cooperation in the OSCE is also difficult, because this organization lacks the necessary funds to implement transboundary cooperation plans for environmental matters; that is, the organization has no financial capacity to offer incentives or to take coercive measures to meet the commitments taken on by its members. The OSCE enables communication with donors and works as a contact platform with other bodies that are able to implement cooperation processes, but it has no a defined or widely accepted identity to manage water resources.

Table 3.5 EUWI current activities

Instruments/frameworks	AM	AZ	GE	KG	KZ	MD	UA	RUS	TJ	TM
New water legislation/strategy on IWRM		●	●	●	●	●	●		●	●
Alignment with the EU Water Framework Directive, and EU Directives on Urban Wastewater or Floods	●	●	●	●	●	●	●		●	●
Work under the UNECE-WHO Protocal on Water and Health	●		●	●		●	●		●	
Climate change adaptation					●	●				
Institutional framework (e.g. River Basin Councils)			●	●		●	●			
Sustainable business models dor WSS systems	●				●	●	●			
Economic instruments for water management						●		●		
Work under the UNECE Water Convention		●	●	●	●		●		●	●

Source: OECD. www.oecd.org/media/oecdorg/directorates/environmentdirectorate/carousel40/EUWI-current-activities-web-700x262.png

NATO, Security, Environment and the Speak Act

After the end of the Cold War, NATO suggested to its former enemies in the Warsaw Pact that they create a framework for cooperation to build confidence. This gave rise to the Partnership for Peace (PfP), created as an instrument for cooperation between NATO and its partner countries in 1994. PfP is a key instrument for collective security. The Partnership for Peace (PfP) is a programme of practical bilateral cooperation between individual Euro-Atlantic partner countries and NATO. It allows partners to build up an individual relationship with NATO, choosing their own priorities for cooperation.

All five Central Asian countries were early participants in the North Atlantic Cooperation Council (NACC)—a forum for dialogue established by the Alliance in December 1991 as a first step in reaching out beyond the East—West divide to former Warsaw Pact members. This body was later replaced by the Euro-Atlantic Partnership Council (EAPC) in 1997. Four out of the five CA countries quickly took advantage of the opportunities offered by the Partnership for Peace, joining this major programme of practical bilateral cooperation shortly after its launch in 1994 (Tajikistan joined later, in 2002). At the Istanbul Summit of 2004, Allied leaders decided to form a partnership with Central Asia, as well as the Caucasus, a priority for the Alliance. Each Central Asian country's relations with NATO has evolved differently, as individual partners are free to choose how and in which areas they wish to cooperate with the organization.

The security concepts included in the North Atlantic Treaty do not mention the links between security and the preservation of the environment or water resources. NATO's priority was coping with a possible attack from the Soviet bloc. This changed after the dissolution of the Soviet Union, because the organization began to consider that the future challenges and risks would be of quite a different nature, as it established in its strategic concept of 1991: "The security challenges and risks which NATO faces are different in nature from what they were in the past. The threat of a simultaneous, full-scale attack on all of NATO's European fronts has effectively been removed and thus no longer provides the focus for Allied strategy. In contrast with the predominant threat of the past, the risks to Allied security that remain are multi-faceted in nature and multi-directional, which makes them hard to predict and assess."

And with regard to the links between environment and security, the Alliance's New Strategic Concept (1991) established that: "It is now possible to draw all the consequences from the fact that security and stability have political, economic, social, and environmental elements as well as the indispensable defense dimension. Managing the diversity of challenges facing the Alliance requires a broad approach to security."

Once the Soviet threat to Europe had faded, the organization focused on new objects and contexts to be securitized, so it became necessary to introduce a series of changes in its strategic concepts, based on the new multidimensionality of the threats and risks, but it was also necessary to look for new partners who might provide the organization with new discourses, rules and values in order to face the new threats, environmental ones among them.

In its Part 3, the strategic concept of 1999 established that: "The Alliance is committed to a broad approach to security, which recognizes the importance of political, economic, social and environmental factors in addition to the indispensable defense dimension.... The United Nations (UN), the Organization for Security and Cooperation in Europe (OSCE), the European Union (EU) and the Western European Union (WEU) have made distinctive contributions to Euro-Atlantic security and stability. Mutually reinforcing organizations have become a central feature of the security environment."

But there was no reference to water resources as an object to securitize until the strategic concept of 2010: "Key environmental and resource constraints, including health risks, climate change, water scarcity and increasing energy needs will further shape the future security environment in areas of concern to NATO and have the potential to significantly affect NATO planning and operations."[20]

Water is thus considered a strategic resource and NATO will see it as a possible conflict and action scene.

SECURITIZATION PROCESS AND DISPOSITIF

The water securitization process within the framework of NATO is related to the activities of the Science for Peace and Security (SPS) Programme. The Programme is a policy tool that enhances cooperation and dialogue with all partners, based on scientific research, innovation and knowledge exchange. It provides funding, expert advice and support to security-relevant activities jointly developed by a NATO member and partner country) and its cooperation relations with United Nations bodies, the OSCE, and the EU. In fact, part of NATO's environmental cooperation with Central Asia is carried out through programs shared with UNECE, UNEP, UE, OSCE and the Environment and Security Initiative (ENVSEC). In 2004, NATO became an associated member of the Initiative, and since then has coordinated with ENVSEC its environmental activities that are aimed at enhancing security in vulnerable regions, and supports selected ENVSEC projects that are in line with NATO's geographical and thematic priorities implemented within the designated Partnership for Peace program.

The Science for Peace and Security (SPS) Programmes in Central Asia related to the environment and water resources are related to the section "Cooperation in science and environmental questions":

1. The creation of a database in a predictive model regarding the Amu Darya river watercourse.
2. The Karakalpak University in Nukus, Uzbekistan, has been supplied with Global Positioning System (GPS) equipment via satellite to gather data about the Amu Darya river in order to improve river management.
3. *Artemia* production. A group of American, Belgian and Uzbek scientists are working in a project in the Aral Sea delta to produce a microscopic organism, *Artemia*, appearing in highly saline environments and used as an additive in fish farming, with the aim of taking advantage commercially of the progressive salinization of the delta.
4. Surveys related to drinking water safety (Microbiological Safety of Drinking Water in Uzbekistan and Kyrgyz Republic; North Atlantic Treaty Organisation 2014) and cross-border pollution of water (Assessing Transboundary Water Pollution in Central Asia; North Atlantic Treaty Organisation 2014).
5. Mélange Uzbekistan—the conversion of rocket fuel oxidizer. The NATO Science for Peace and Security Committee agreed to carry out and fund a Science for Peace project (2009–2010) using the mobile NATO plant that transformed mélange into an environmentally friendly liquid.
6. Prevention of landslide dam disasters in the Tien Shan. This project, completed in 2009, aimed at mitigating the damage caused by landslides, through the use of regional mapping, field investigations and 3D-modelling of the hazards and resulting risk scenarios.

7. Assessing transboundary water pollution in Central Asia. This multi-year research project, concluded at the end of 2013, brought together Uzbekistan, Tajikistan, Kyrgyzstan and Kazakhstan to conduct an in-depth study of contaminants in the basin of the Syr Darya river.

8. Uranium extraction and environmental security in Central Asian republics. Subject matter experts from the Central Asian countries cooperated in a multi-year project (2006–2012) to manage uranium industry wastes in order to prevent adverse effects on the health of local populations and on the environment.

ENVSEC PROJECTS IMPLEMENTED BY NATO IN CENTRAL ASIA

Biotechnical exploration of Uzbek saline water reserves using halotolerant microalgae (2006–2010)

Microbiological safety of drinking water in Uzbekistan and Kyrgyz Republic (2007–2011)

Environment security issues arising from the legacy of uranium extraction in Central Asia (2005–2011). Using stable isotopes, passive organic samples and modelling to assess environmental security in Khorezm, Uzbekistan (2006–2010)

New technologies of seismic resistant construction (2006–2010)

Assessment and mitigation of seismic risk in Tashkent, Uzbekistan and Bishkek, Kyrgyz Republic (2000–2003)

Sustainable development of ecology, and land and water use, through the implementation of a GIS and remote sensing center in Karakalpakstan Uzbekistan (2000–2004)

SEMIRAD I—Investigation of the radiological situation in the Sarzhal Region of the Semipalatinsk nuclear test site (2000–2003)

SEMIRAD II—Investigation of the radiological situation in the Sarzhal Region of the Semipalatinsk nuclear test site (2004–2007) Study of radioactive waste disposal sites in Turkmenistan (2001–2009)

Integrated water resources management for wetlands restoration in the Aral Sea basin (2004–2009)

The overall objective of the project was to propose a system of models, GIS and engineering tools for civil infrastructure and a pre-feasibility study meeting the principal needs for integrated water resources management in the Syr Darya Delta

Geo-environmental security of the Toktogul hydroelectric power station (2008–2012)

None of these activities is therefore related to multilateral cooperation in the Amu Darya basin, but rather to technical questions that are far from the values and principles of the human right to water.

NOTES

1. Article 12 of The International Covenant on Economic, Social and Cultural Rights (ICESCR recognizes the right to the "enjoyment of the highest attainable standard of physical and mental health". The steps to be taken by Parties towards the realization of this right shall include, inter alia, "the improvement of all aspects of environmental and industrial hygiene" (Article 12.2 (b)).

2. Committee on Economic, Social and Cultural Rights, 22nd Sess., General Comment 14, 14, 36, U.N. Doc. E/C.12/2000/4 (11 August 2000).

3. John Knox is the first independent expert on human rights and the environment. He began working on the mandate on 1 August 2012. OHCHR. http://www.ohchr.org/EN/Issues/Environment/IEEnvironment/Pages/IEenvironmentIndex.aspx.

4. Zaharchenko, Tatiana R.: *Environmental Policy in the Soviet Union.* (UCDAVIS. University of California). http://environs.law.ucdavis.edu/volumes/14/1/articles/zaharchenko.pdf.

5. EJF (2012): *The true costs of cotton: cotton production and water insecurity.* Environmental Justice Foundation, London.

6. Temperature rise development (it will continue in years to come): Uzbekistan (1950–2005): 0.290C; Kazakhstan (1936–2005): 0.260C; Turkmenistan (1961–1995): 0.180C; Tajikistan (1940–2005): 0.100C; Kyrgyzstan (1883–2005): 0.080C.

7. Eurasian Development Bank. *Executive Board of the International Fund for Saving the Aral Sea. Regional Center of Hydrogeology. Impact of Climate Change to Water Resources in Central Asia.* (Consolidated Report) http://www.cawater-info.net/library/eng/ifas/impact_climate_change_en.pdf.

8. *Regional Water Intelligence Report. Central Asia Baseline Report.* By: Jakob Granit, Anders Jägerskog. Rebecca Löfgren, Andy Bullock, George de Gooijer, Stuart Pettigrew and Andreas Lindström. Stockholm, March 2010. http://www.watergovernance.org/documents/WGF/Reports/Paper-15_RWIR_Aral_Sea.pdf, p. 8.

9. Abdulkasimov, H. P., Alibekova, A. V. and Vakhabov, A. V. (2003): *Desertification problems in Central Asia and its regional strategic development.* Abstracts, NATO Advanced Research Workshop. Samarkand, 11–14 June. p. 4. (In Russian). Water-related Problems of Central Asia: Some Results of the (GIWA) International Water. Assessment programme Igor Vasilievich Severskly. http://www.unep.org/dewa/giwa/publications/articles/ambio/article_7.pdf pp. 7–8.

10. State Committee for Land Management of the Republic of Tajikistan. Resume National Report or the Republic of Tajikistan to Combat Desertification. Dushanbe—2006. www.cawater-info.net/library/eng/tj/st_land_com.pdf.

Stockholm Convention on Persistent Organic Pollutants (POPs). National Implementation Plan of the Republic of Kazakhstan on the obligations under the Stockholm Convention on Persistent Organic Pollutants. 2009. http://www.pops.int/documents/implementationf.
UNCCD. Implementation in Turkmenistan. National Report of the Republic of Uzbekistan on the Implementation to Combat Desertification. 2010 (CCD). www.unccd.int/ ... /reports/ ... /national/ ... /turkmenistan-summary-eng. pdf. The Ministry of Environmental Protection of the Republic of Kazakhstan. The Third National Report of the Republic of Kazakhstan on Implementation of the United Nations Convention to Combat Desertification. Astana, 2006. www.cbd.int/doc/world/kz/kz-nr-03-en.pdf.

11. WNN | World Nuclear News. EBRD launches uranium mining legacy fund. http://www.world-nuclear-news.org/UF-EBRD-launches-uranium-mining-legacy-fund-1806157.html.

12. UNECE. *Environmental Policy Under Scrutiny in Kyrgyzstan. A backlash surrounding a gold mine is placing the European Bank for Reconstruction and Development under scrutiny.* By Ryskeldi Satke and Franco Galdini. http://aarhusclearinghouse.unece.org/news/1000828/?year=2014.   EBRD's November 17, 2014.

13. Anders Aslund, Martha Brill Olcott, Sherman W. Garnett. *Getting It Wrong: Regional Cooperation and the Commonwealth of Independent States,* Carnegie Endowment for International Peace (February 2000).

14. Data from Ronald B. Mitchell. 2002–2015. International Environmental Agreements Database Project (Version 2014.3). Available at: http://iea.uoregon.edu/. Date accessed: 1 September 2015.

15. Evangelista, M. (1986): "The New Soviet Approach to Security". World Policy Journal, Vol. 3, No. 4, pp. 561–599. Published by: The MIT Press and the World Policy Institute. Stable URL: http://www.jstor.org/stable/40209031. Accessed: 20 March 2011 20:41.

16. By Wu Jiao and Li Xiaokun (China Daily). Updated: 2010-06-12 08:22 http://www.chinadaily.com.cn/cndy/2010-06/12/content_9968565.htm.

17. IOSR Journal of Humanities and Social Science (IOSR-JHSS) Volume 19, Issue 1, Ver. III (Jan. 2014), pp. 91–94. e-ISSN: 2279-0837, p-ISSN: 2279-0845. www.iosrjournals.org www.iosrjournals.org. p. 91.

18. Ibid.

19. OSCE Yearbook 2009. Yearbook on the Organization for Security and Co-operation in Europe (OSCE), Institute for Peace Research and Security Policy at the University of Hamburg /IFSH (ed.), p. 348.

20. NATO strategic concept. http://www.nato.int/lisbon2010/strategic-con cept-2010-eng.pdf. (Part III).

CHAPTER 4

Economic Security, Water Resources and International Security Organizations (NATO, OSCE, CSTO, SCO)

Abstract Water is an irreplaceable resource involved in economic activities that are basic to human development, such as agricultural, industrial and energy production. The quality and amount of water available in fluvial basins, transboundary or not, along their watercourses depends mainly on the existing agricultural and energy—industrial model, sectors that are subject to the management of water resources. The protection and implementation of the Human Right to Water and Sanitation (HRWS) ensure the economic possibilities of all the citizens and communities involved, not only of a minority. In this chapter, the author establishes the economic relevance of water in the Amu Darya basin and its links with the vulnerability and human security of its populations, as well as the competence between the different uses of water in the region (water—energy—economy). Finally, the author analyzes the cooperative response to economic problems in the area and evaluates security policies of international security organizations regarding economic security.

Keywords Poverty · Food security · Poverty energy · Cotton · Slavery

© The Author(s) 2017
M.Á. Pérez Martín, *Security and Human Right to Water in Central Asia*, Security, Development and Human Rights in East Asia,
DOI 10.1057/978-1-137-54005-8_4

INTRODUCTION

Poverty is the common denominator of economic insecurity. It is seen by many as the outstanding economic and social problem in the world (Nef 1999). Human rights are based on principles of dignity and freedom, and both are severely compromised when human beings cannot meet their fundamental needs. Economic and social rights guarantee that every person should be afforded conditions under which they are able to meet their needs. In particular, economic and social rights include: the right to education, food and housing, the right to social security, and the right to work. Food security, employment, access to housing, hygiene, access to education or any other production system ultimately depend on the access to water and its quality.

Water is an irreplaceable resource involved in economic activities that are basic to human development and the process of emancipation, such as agricultural, industrial and energy production. The amount and quality of water available in fluvial basins, transboundary or not, along their watercourses depends mainly on the existing agricultural and energy-industrial model, sectors which at the same time are subject to the management of water resources. The protection and implementation of HRWS ensures the economic possibilities of all the citizens and communities involved, not only of a minority.

In the next section we shall establish the poverty rates in Central Asia, the economic relevance of water in the Amu Darya basin and its links with the vulnerability and human security of its populations, and the competence between the different uses of water in the region.

GROWTH AND POVERTY

In general terms, Central Asian countries did not stopped growing between 2008 and 2015 (Table 4.1).

The region's economic growth in recent decades was generally linked to the exports of raw materials. The highest growth took place in those states whose main source of income was the export of energy raw materials. Kazakhstan, in 2009, became the thirteenth in the list of countries by oil exports, with more than 1 million barrels per day, and Turkmenistan was rated the sixth among world gas exporting countries, according to Energy International Agency EIA data. On the other hand, Kyrgyzstan and Tajikistan,

Table 4.1 CCA: Selected economic indicators

	Average projections	2009	2010	2011	2012	2013	2014	2015
Real GDP growth (Annual change; %)								
Kazakhstan	9.9	3.7	6.8	6.8	5.7	6.6	6.2	6.4
Kyrgyzstan	9.4	1.2	7.3	7.5	5.0	6.0	5.7	6.1
Tajikistan	4.9	2.9	-0.5	6.0	-0.9	10.5	4.4	4.9
Turkmenistan	8.6	3.9	6.5	7.4	7.5	7.4	6.2	5.7
Uzbekistan	15.1	6.1	9.2	14.7	11.1	10.2	10.7	12.5
	6.3	8.1	8.5	8.3	8.2	8.0	7.00	6.5

Source: National authorities; and IMF staff estimates and projections. Regional Economic Outlook Update. Middle East and Central Asia Department.
https://www.imf.org/external/pubs/ft/reo/2014/mcd/eng/pdf/cca0514.pdf

lacking large sources of hydrocarbons, and Uzbekistan, whose significant gas reserves are now mainly devoted to domestic use, have not grown as much.

The Tajik economic growth is basically connected to a certain degree of social stability thanks to the end of the civil war, money from immigration, and its resumption of cotton exports. In the case of Kyrgyzstan, the growth is linked to the export of gold, and in Uzbekistan to the high prices of cotton and increasing exports of gold and gas. In recent decades, growth in Central Asia has generally been linked to price increases of raw materials, as can be seen in the next table (Table 4.2)

Has this growth had any influence on Central Asian citizens' employment and standard of living?

In spite of the economic growth, in some cases poverty has continued to increase in the region over recent decades. According to World Bank data, the percentage of people living on less than 4.5 US$ dollars a day has gradually increased in recent decades in all Central Asian countries, apart from Kazakhstan. Let us take a look at the next table (Table 4.3)

–In recent decades there has been an increasing amount of inequality, and the growth of wealth does not seem to have benefited the whole population, as shown by Gini coefficients (Table 4.4)

The unemployment rates in the region have been relatively stable through recent decades in Turkmenistan, Uzbekistan and Tajikistan, at slightly over 10 %, and Kazakhstan is the country with the highest (though not spectacular) levels of job creation, reducing its unemployment rate from 8.2 % in 2005 to 5.3 % in 2013.

In contrast, immigration rates, the informal economy and criminal activities have increased considerably in the region. In 2010, between 2.5 and 4 million people from Central Asia were working abroad, mainly

Table 4.2   Prices of oil, cotton and gold

	2000	2001	2002	2003	2004	2005	2006–2009	2010–2015
Oil (a)	28.5	24.5	25	28.8	38.3	53.0	45.50	40.45
Cotton (b)	59.2	48	46.2	63.3	63.6	63.3	60	58
Gold (c)	279	271	310	363	421	345	310	320

Source: OEF, EIU, IMF WEO, and staff estimates. This table was first published as part of an article by the Asian Development Bank (www.adb.org), "Central Asia's Economy: Mapping Future Prospects to 2015", Malcolm Dowling Ganeshan Wignaraja Silk Road Paper published by the Central Asia-Caucasus Institute & Silk Road Studies Program 2006. http://www.adb.org/sites/default/files/publication/28200/wp080.pdf.

Table 4.3 Income poverty trends (1981–2010, PPP$4.30/day threshold)

	1981	1984	1987	1990	1993	1996	1999	2002	2005	2008	2010
Kazakhstan	n.c.	n.c.	n.c.	n.c.	62 %	59 %	57 %	66 %	50 %	29 %	30 %
Kyrgyzstan	n.c.	n.c.	n.c.	n.c.	58 %	84 %	85 %	95 %	83 %	62 %	70 %
Tajikistan	n.c.	n.c.	n.c.	n.c.	92 %	99 %	98 %	95 %	90 %	87 %	79 %
Turkmenistan	n.c.	n.c.	n.c.	n.c.	n.c.	n.c.	n.c.	n.c.	n.c.	n.c.	n.c.
Uzbekistan	n.c.	n.c.	n.c.	n.c.	n.c.	n.c.	n.c.	n.c.	n.c.	n.c.	n.c.

Source: World Bank POVALNET database. Pre-2002 poverty rates for most countries are computed using 2005 ICP PPP exchange rates. Poverty rates for other years computed using 2008 ICP PPP exchange rates

n.c. = Data are available, but not credible.

Table 4.4 Gini coefficients of income inequality from the SWIID database

	1981	1984	1987	1990	1993	1996	1999	2002	2005	2008	2009	2010	2011
Kazakhstan	0.228	0.231	0.233	0.234	0.312	0.328	0.327	0.321	0.345	0.301	0.286	0.282	0.282
Kyrgyzstan	0.223	0.222	0.220	0.226	0.411	0.447	0.356	0.306	0.337	0.367	0.356	0.360	0.350
Tajikistan	0.222	0.244	0.270	0.296	0.302	0.308	0.314	0.324	0.327	0.316	0.314		
Turkmenistan	0.231	0.219	0.215	0.240	0.291	0.361	0.410	0.419	0.405				
Uzbekistan	0.223	0.229	0.238	0.248	0.336	0.367	0.371	0.337	0.352				

Source: Fuente; UNDP, Poverty, inequality, vulnerability in the transition countries economies of Europe and Central Asia. 2014, page 45

in Russia, and the countries with the largest flows were Tajikistan, between 26.4 % and 43.9 % of its active population; Kyrgyzstan, 13.1 %–28.6 %; Turkmenistan, 10.6 %–15.9 %; Uzbekistan, 10.3 %–12.9 %; and Kazakhstan, 4.1 %–5.8 %.[1]

Studies indicate that the Central Asian population became much more active in terms of migration in the 2000s, and migration flows were joined by new socio-demographic population groups. Residents of rural areas, small settlements, women and young people now take a more active part in migration plantations in the Volgograd Region. In that case, de facto labor migration in the form of the exploitation of child labor was labeled 'summer study internships for high school students'.[2]

According to approximate estimates provided by experts, currently the number of labor migrants who live in conditions similar to slavery is around 600,000 persons, or 20 % of all migrant workers in Russia. According to estimates provided by the Chair of the Association of Russian Lawyers for Human Rights Year, Evgeny Arkhipov, the cost of one person coming from Central Asia varies between 300 and 500 US$ on the black market, and "one can order a slave through criminal organizations and foremen of migrant workers".[3]

Migrants turn into slaves primarily because of debts: they have to pay 2,000 roubles to criminal organizations simply to be able to go to Russia and return home. The US State Department Trafficking in Persons Report 2012 included Russia in the group of countries where "the absolute number of victims of severe forms of trafficking is very significant or is significantly increasing". The Center for Migration Research (CMR) estimates that there could be some 1–1.5 million people in the position of being de facto slaves, According to estimates by non-governmental organizations (NGOs), in Russia around 4 million migrants are victims of labor exploitation close to slavery.

According to the 120-country world rankings published by the World Bank in 2010, the informal economy in Tajikistan, Kyrgyzstan and Kazakhstan represents more than 40 % of gross domestic product (GDP); there are no data available for Uzbekistan and Turkmenistan.[4]

Criminal activities, especially those linked to drug trafficking—in 2009 Afghanistan exported almost 95 tons of heroin via Central Asian countries, with a final market value of 13 trillion US dollars, one of the main routes in the region after Pakistan (150 million US dollars) and Iran (105 million US dollars). Most of this trafficking is carried out across the Afghanistan border with Turkmenistan, Uzbekistan and Tajikistan, drawn by the Amu Darya river watercourse and its tributaries; the pass of this river boundary of more than 2,000 km thus marks the starting line for international heroin trading, and is a key point within the trafficking networks.

The increasing degree of inequality might be caused by several complex reasons, different in each country, but some general factors could usefully be mentioned in order to explain the origins of the unequal wealth distribution in the region[5]:

1. The slow pace of price and trade liberalization in some of these countries, especially Uzbekistan and Turkmenistan, has produced an "elite of privileged traders" thanks to favors from the political and administrative elite.

2. The privatization of companies has been carried out without the necessary transparency (benefiting the old nomenklatura members) and at very low prices, especially affecting companies with the highest opportunities for generating benefits, undercapitalizing the state's capacity to create wealth or its ability or will to control the big firms' economic activity (taxes).

3. The high inflation rate in the 1990s wiped out most workers' savings and salaries. Since then, there have been drastic cuts in health, education and pensions because of the lack of public funding and lack of transparency in the acquisition of social benefits. Salaries have been kept low, especially in companies, bodies and institutions that depend on the government.

In this context, the sectors affected most negatively in terms of social inequalities have been the unemployed, part-time workers, public workers (education, health, science and the arts) and their families, farming day laborers, small traders and their families, especially in rural areas, young unemployed people without work experience, pensioners, refugees and displaced persons.

The situation of women ostensibly worsened after independence, and the economic crisis has affected the female workforce irreversibly. Education- and health-related cuts have in particular affected the situation of women in the region:

1. The main niches of women's employment—health, education and social services—have been sectors particularly affected by spending cuts in new Central Asian states, forcing women to carry out

housekeeping tasks, do low-qualification jobs (domestic service, agriculture), or to work in the informal economy sectors.

2. The cuts have had a direct influence over social services, and baby-sitting and childcare, so women have found themselves forced to leave their jobs to take care of their families.

In the new social post-Soviet order, there is another factor sending women back to household tasks or weakening their degree of participation in public life: the renaissance of traditional values from before the Soviet era, when the main role of women was to become mothers and carry out their domestic duties; in Tajikistan, for example, Women's Day has been transformed into Mother's Day. Despite that, however, Central Asian countries have signed most of the international agreements that support women's rights.

Inequality between men and women shows up not only in terms of access to employment, but also in their salaries: a woman earns between 20 % to 40 % less than a man doing a similar job. Women are not protected against discrimination by a specific legislation or by institutions that could enable their access to justice if their work rights were violated.[6]

ECONOMIC RELEVANCE IN THE AMU DARYA BASIN

Introduction

The Amu Darya basin has a population of some 50 million people. Water from this basin (except in Afghanistan) is used mainly in agriculture: approximately 5 million hectares are under cultivation (cotton, wheat, pastures, fish hatcheries, fruit and vegetables, etc.). The rest is dedicated to electric power production and to Uzbek industrial complexes (metallurgy, extraction activities, gold, uranium, gas in Navoi, Bujara and Samarkand), Tajik (aluminum and cement industries in Dushanbe and Tursunzade), and Turkmen (petrochemical and gas complexes) (UNEP_GRIDA Environment and Security in the Amu Darya River Basin 2011).

Agriculture and Food Security

In the Amu Darya basin more than 5 million hectares are cultivated, and more than 3 million in the basin of its neighboring river, the Syr Darya.[7]

Agriculture consumes 91 % of the available water in the region,[8] and most of the population in Central Asia works in that sector: 40.6 % (*FAO statistical yearbook 2014: Europe and Central Asia food and agriculture*). However, the countries of this area are not able to meet their own food demands. Food insecurity affects all Central Asian states, to varying degrees. Tajikistan is able to cover only 31 % of its food consumption needs, while Kyrgyzstan, Uzbekistan and Turkmenistan cover around 50 % of theirs.[9]

The table shows the data for undernourished people in the region (Table 4.5).

Since the 1990s, according to the Food and Agriculture Organization (FAO) figures, undernourished people and the prevalence of undernourishment have gradually decreased in all the region's countries, though data for Kazakhstan and Turkmenistan are missing. However, the total number of

Table 4.5   Progress towards achieving MDG and WFS hunger targets in the CCA sub-region, number of undernourished (millions) and prevalence of undernourished (%)

		1990–1992**	2000–2002	2005–2007	2008–2010	2011–2013	*Progress*
Armenia	mln.	0.8	0.6	0.2	ns	ns	+
	%	24.0	20.2	5.3	<5	<5	+
Azerbaijan	mln.	1.8	0.8	ns	ns	ns	+
	%	23.8	10.1	<5	<5	<5	+
Kazakhstan	mln.	ns	1.2	ns	ns	ns	+
	%	<5	8.4	<5	<5	<5	+
Kyrgyzstan	mln.	0.8	0.9	0.5	0.5	0.3	+
	%	17.7	17.6	9.7	9.3	5.9	+
Tajikistan	mln.	1.7	2.6	2.3	2.5	2.1	–
	%	30.3	42.1	34.9	37.1	30.2	–
Turkmenistan	mln.	0.4	0.4	0.3	ns	ns	+
	%	9.2	8.4	5.7	<5	<5	+
Uzbekistan	mln.	ns	3.9	2.5	2.2	1.6	–
	%	<5	15.7	9.7	8.1	5.7	–
Central Asia and Caucasus	mln.	9.7	11.6	7.3	7.0	5.5	+

Source: FAO (2013)

**Estimates of food security indicators for 1990–1992 are missing for the CCA countries from FAO (2013); thus 1992–1994 estimates have been used instead, as we do not anticipate much change between the two periods.

undernourished people is still high: more than four million people, according to the data Table 4.5. While the prevalence rates of undernourished people are high in Tajikistan (30.2 %), Kyrgyzstan (5.9 %) and Uzbekistan (5.7 %), in Afghanistan the figures are much worse. Approximately 2.2 million Afghans live on less than 1,500 kilocalories per day and are considered very severely food insecure. Food insecurity affects nearly 8 million people, with an additional 2.4 million classified as severely food insecure, and 3.1 million moderately food insecure. There are 1.2 million children acutely undernourished, and 500,000 of them younger than five years of age will need treatment for severe acute malnutrition. Malnutrition is an underlying cause in more than one-third of under-fives child deaths in Afghanistan.

One of the critical factors behind this situation of food insecurity has been the excessive preference of the agricultural sector for cotton production, using large amounts of water at the expense of different kinds of crops for human consumption.

Cotton Monoculture in Central Asia

In 1929, the Central Committee of the USSR Communist Party undertook the collectivization of the agricultural means of production. This meant a revolutionary and dramatic change in the socioeconomic patterns in the Central Asian region, traditionally consisting of landowners' farming communities and nomadic shepherds.

These were the main measures of collectivization:

(a) Confiscation of local landowners' estates, which were distributed among farmers. The first collective farmers (*kolkhozy* and *sovkhozy*), known as "*koschi*", appeared in Central Asia.
(b) Expropriation of water rights from the old institutions. Control over water resources to be managed by the local Soviets.
(c) Enforced sedentarization of nomadic tribes.

Collectivization in Central Asia was closely linked to the introduction of cotton monoculture. The expansion and intensification of this crop was started under Russian colonization, but it was definitely imposed when the Bolsheviks came to power.

By the start of the twentieth century, Central Asia had acquired a new significance for the Russian economy. The tsarist government supported cotton production in Turkestan, which it saw as a way of achieving freedom for the textile industry of European Russia from the cotton imports from the USA and India. In the often-quoted words of the Minister of Agriculture in 1912, "Every extra pud of Turkestani wheat [provides] competition for Russian and Siberian wheat; every extra pud of Turkestani cotton [presents] competition to American cotton. Therefore, it is better to give the region imported, even though expensive, bread, [and thus] to free irrigated land in the region for cotton."[10]

Subsequently, the construction of an extensive network of canals allowed the cotton production boom. In 1913, the area used for cotton crops was 441,660 hectares, and 1,022,600 hectares in 1940; two years later, that extension had continued to grow up to two million hectares of the cultivated lands (Spoor 1993, pp. 147–149).

Cotton thus replaced other traditional crops for local consumption in the region such as rice, cereals, fruit, vegetables and so on, and the Soviet republics became dependent on the Russian wheat, with the exception of Kazakhstan, since at that time a large producer of wheat because of the ploughing up and colonization campaign of the Kazakh steppes promoted by Nikita Khrushchev, Premier of the Soviet Union, in 1953.

Cultivation of cotton and its development model in Central Asia has meant, and still means, a high cost in terms of social and environmental resources. Uzbekistan is the best example of this.

UZBEK COTTON

Uzbekistan, the most populous country in the Amu Darya basin, with about 28 million people in 2012 (*Statistical Yearbook 2014: Europe and Central Asia food and Agriculture*), is the largest water consumer in this basin.

The cultivation of cotton in Uzbekistan, as in the rest of the region's countries, consumes considerable resources[11]:

- 1.4 million hectares of land, or 36–37 % of all agricultural land, is used to grow cotton.
- Of the 53.1 billion cubic meters of water consumed annually, 92 % goes for agricultural needs, the lion's share of which is consumed by the cotton sector.

- Agricultural production and processing of agricultural products employs 30–35 % of the working population, more than half in the cotton sector. This does not include people forced to harvest cotton each year.
- The cotton sector consumes a significant portion of the mineral fertilizers, such as ammonium nitrate, ammonium sulphate and urea, produced by the chemical industry, and 290,000 tons of diesel fuel, or more than 30 % of domestic consumption, as well as other resources. This means that one of the biggest polluters in this basin is the cotton sector.

The amount of resources used up by this sector is huge. Let us now analyze how it is managed. Cotton is one of Uzbekistan's major exports, as happens in other countries of the Amu Darya basin such as Turkmenistan and Tajikistan, and constitutes a major source of hard currency for the state. Uzbekistan is the world's sixth largest producer of cotton, and fourth largest exporter, accounting for 5 % of global production of cotton fiber.[12]

Uzbekistan exports approximately one million tons of cotton fiber annually, bringing in revenues of over US$1 billion, depending on the world market, comprising, according to official data, 11 % of total export earnings.[13]

The government continues to exercise extensive control in agriculture, particularly in cotton and wheat production, which are referred to as "centralized" crops. The state maintains ownership of the land, and the right to use land agriculture (other than household plots) is conditional on acceptance of the state's quotas for planting cotton and wheat. The state also provides subsidized inputs; including irrigation. The "state order" system also includes quotas on the production of cotton and wheat, as well as on the area planted.[14]

To ensure quotas are met, the state monitors efforts year-round (Veldwisch and Spoor 2008): leaching is monitored in the winter; planting areas, varieties and dates are determined by the state in the spring; and fertilizer application during the growing season is directed by Ministry of Agriculture and Water Resources (MAWR) officials. During the cotton-growing season, state officials visit farms to determine yield potential, and adjust planning targets and production quotas. Annual planting area is determined by a state plan and, at the local level, planning may extend to determining which fields are used for cotton, wheat or non-centralized crops. While the state and collective farms organized during the Soviet period have largely been privatized since 2006, a large number of farms still rely on centralized Machine Tractor Parks (MTP) for machinery, and MTPs prioritize centralized crops.[15] As Bakhodyr Muradov and Alisher Ilkhamov say: "The cotton sector... it is still controlled by an administrative command system of management, despite de-collectivization and the legal conversion of the majority of collective and state farms into private farms, which numbered 66,000 by the beginning of 2013"[16]

Though legally the farms are private, in fact the farmers rent land from the government and are not free to make decisions about the use of the land allotted to them, to choose the crops they plant, or to select suppliers of inputs, or buyers for their products. Each year the local authorities and farmers receive mandatory quotas for cotton and grain production. For failure to meet the quotas, local *hokims* (Uzbek governors of administrative regions) risk losing their jobs, and farmers are subject to a range of economic and administrative sanctions, including criminal prosecution.[17]

To understand the reasons for this state of affairs, it is essential to examine the financial flows that have been established in and around the sector, and to expose the interests and the real benefits received by different participants in the system, starting first with the government.

A comprehensive and multifaceted analysis must take into account the diverse nature of the cotton sector, which is not limited to the cultivation of cotton and the extraction of its byproducts, such as cotton fiber, cottonseed oil, etc. The participants in the sector include[18]:

1. Suppliers of production inputs, including the state joint-stock company Uzkhimprom, which produces mineral fertilizers and crop protection chemicals; the national oil and gas holding company Uzbekneftegaz, which supplies fuel; agro-universities and research institutes; seed producers; agricultural equipment producers; leasing companies; banks; firms providing mechanization services; agencies responsible for irrigation, and soil amelioration.
2. Cotton farmers.
3. Cotton processing enterprises: Khlopkoprom (also known as Uzkhlopkoprom), the state-controlled agency supervising and operating cotton procurement and cotton gins; companies involved in the processing of seeds and production of cottonseed oil; and textile and knitting industries.
4. Trading companies and companies providing transportation services.

However, the cotton business supports the exploitation of women and children by most of the individual producers, who are also employed in collective farms: these producers find themselves forced to use women and children as workers in order to compete with the collective farms, whose workforces also include women and children too. These sectors of the population so carry out the hardest and worst paid jobs in both kinds of farms – collective and family run –, a fact that leads them to a lack of different working and educational opportunities. Despite the difficulties and data concealment by Central Asian governments with regard to this situation, according to The United Nations Children's Fund (UNICEF)'s estimations in the year 2000 in Uzbekistan, where this practice seems to be more widespread, 22 % of the Uzbek children worked part-time in cotton-harvesting (for 3 months), underfed and in deplorable health conditions, with a high risk of contracting diseases because of the chemical fertilizers used in cotton cultivation. Near 200,000 in the Ferghana Valley and 60,000 in Jizzak are employed in these sorts of jobs. Central Asian agriculture productivity is thus partially based on the exploitation of women and children.

The cotton sector is therefore economically and socially a very important one, but its system of water management is unsustainable in economic, environmental and human terms, for several reasons:

1. Excessive water consumption, as in the case of Uzbekistan's irrigation system, where 50–80 % of water used for agricultural irrigation is lost. Only 25–35 % of what reaches the crops is used efficiently.[19]
2. Environmentally unsustainable irrigation techniques make the soil unproductive. The problem of salinization is especially acute in Uzbekistan, where over 50 % of the irrigated land is affected in varying degrees as a result of inappropriate irrigation practices. Salinization is one of the country's most serious environmental problems, the UN Food and Agriculture Organization (FAO) maintains.[20]

The consequence of this inefficiency in water management is that a labour force close to slavery ultimately pays a high price; a former senior provincial official from rural Uzbekistan said: "[Farmers] are told they have to grow cotton, and the way they water the fields of cotton is very oldfashioned. They should use new modern methods to do it, but [the government] does not want to spend money. They could buy cotton-picking machines, but it is cheaper for them to use children and the people's labour for cotton picking."[21]

WATER AND ENERGY

Water also plays a fundamental role in the region's generation of energy. More than 80 % of the electricity produced in Uzbekistan and Turkmenistan comes from thermal plants whose raw material is gas and using great amounts of water; in Tajikistan and Kyrgyzstan also, more than 80 % of energy production is hydroelectric.

In the Amu Darya basin, the potentialities in terms of energy supply (water and hydrocarbons) for the population are spectacular. Turkmenistan is the sixth-largest natural gas reserve holder in the world and was the second-largest dry natural gas producer in Eurasia, behind Russia, in 2012. However, the hydrocarbon-rich country lacks sufficient pipeline infrastructure to export greater volumes of hydrocarbons.[22]

Uzbekistan was the third-largest natural gas producer in Eurasia, behind Russia and Turkmenistan, in 2012. Uzbekistan had 594 million barrels of proven crude oil reserves as of January 2014, according to *Oil & Gas Journal*. In 2013, total crude oil and other liquid production was about 102,000 barrels per day (bbl/d), of which 30 % came from natural gas plant liquids. Roughly 60 % of all known oil and natural gas fields are located in the Bukhara-Khiva region. The region is the source of approximately 70 % of the country's oil production. Uzbekistan's natural gas transmission and distribution system allows for trade with Russia, Kazakhstan and Kyrgyzstan. Uzbekistan also serves as a transit country for natural gas flowing from Turkmenistan to Russia and China. In addition, two new natural gas pipelines, Gazli-Kagan and Gazli-Nukus, were built to connect the Ustyurt and Bukhara-Khiva region with the existing system.[23]

A December 2011 study by the US Geological Survey (USGS) estimated that Afghanistan is endowed with potentially exploitable reserves of 946 billion barrels of oil and 7 trillion cubic feet of natural gas, within the Afghan—Tajik Basin. Afghanistan could, for example, use its domestic gas supplies for electricity generation to lower its import dependence on gas-generated electricity from Uzbekistan and

Turkmenistan, and use its own domestic oil supplies to lower its dependence on imported diesel fuel.

In spite of the abundance of energy resources, for the population of certain areas around the Amu Darya access to energy is very deficient and data about this are hard to find. However, there is some valuable news on this question, as in Samarkand (Uzbekistan), the Asian Development Bank (ADB), in collaboration with the Uzbekistan Welfare Improvement Strategy, is carrying out a plan devised to provide more than 700,000 people with an adequate electric supply. According to the ADB, "The population of the Samarkand district usually has energy supply for 2 hours per day in winter and an average of 16–20 hours per day in the summer". Poor power supply impacts negatively on (i) heating; (ii) health care services; and (iii) potable water. This situation leads to a high rate of acute intestinal infections and cold-related illnesses. This situation of energy poverty also affects education, according to the same report: "The education system in the project area has suffered greatly from power cuts and voltage dips and spikes, impacting upon the ability to use electrical equipment as well as lighting."[24]

The situation in Tajikistan is no better. Approximately 70 % of the Tajik people suffer from extensive shortages of electricity during the winter. Shortages occur in winter when the demand is high but water flows are low, whereas in summer there has been significant under-utilized power surpluses. The shortages, estimated at about 2,700 GWh, about a quarter of the winter electricity demand, impose economic losses estimated at over US$200 million per annum, or 3 % of GDP. In addition to the financial costs of inadequate electricity, the Tajik people also suffer the social costs, including indoor air pollution from burning wood and coal in homes, and the health impacts from extreme winters. Electricity shortages in rural areas affect the quality of social service delivery. Some schools and medical facilities face the same electricity rationing as residential areas, which affects their functioning. They can only operate during daylight hours. Densely populated areas have a special electricity line for social buildings (a so-called "red line"), which supplies unlimited electricity during the heating season. However, it is common for private houses or small shops to connect illegally to the red line, resulting in social buildings receiving less energy than they need. In the winter of 2008, no heating systems were operating in 26 % of Tajikistan's schools and health centers. The electricity shortages increased considerably in 2009, when Tajikistan's energy trade with neighboring countries through the Central Asia Power System (CAPS) stopped; combined with the continued ageing of Tajikistan's power generation assets, the situation has become worse. Imports of piped natural gas from Uzbekistan were stopped in 2012 as the two countries could not reach an agreement on the price. As a consequence of all this, severe load-shedding occurs in winter, and customers receive electricity only three to seven hours per day in every region except in the capital Dushanbe and in Gorno-Badakhshan Autonomous Province (GBAO). In Afghanistan, just 20 % of the population is connected to the public power grid, and about 60 % of all connections are concentrated in the capital city of Kabul.[25]

Only around 9 % of the rural population, which constitutes more than 70 % of Afghans, has access to electricity and instead relies on self-supplied energy sources, using wood, dung and other biomass for fuel. In contrast, the remaining fewer than 30 % of Afghans who live in urban areas have electrification coverage estimated to be at more than 70 %, including electricity generated from privately owned diesel-power generators.[26]

Poor access to electricity has been identified by the World Bank's Investment Climate Survey as the number one obstacle to investment and business development in Afghanistan.[27] The existence of large energy poverty areas in the Amu Darya basin are related to different factors:

1. The infrastructures of electric energy generation and transportation inherited from the Soviet Union are inadequately maintained, and this has led to a poor supply and permanent energy losses, as the figure shows (Fig. 4.1).

   %Kyrgyzstan wastes 30 % of the energy it generates; Uzbekistan, 20 %, and Kazakhstan, 10 %; there are no figures from Tajikistan, Turkmenistan or Afghanistan.

2. Industrial complexes in this environment are inefficient and consume an excessive amount of energy from electricity. For example, Uzbekistan is the most energy inefficient country in Europe

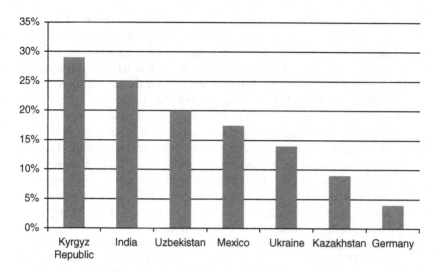

Fig. 4.1 Technical and commercial losses compared with other countries (2010) % of net generation

Source: World Bank team based on WDI and public data from national utilities or energy regulators 2013. International Bank for Reconstruction and Development/The World Bank. Uzbekistan Energy/Power Sector Issues. p. 21.

and Central Asia (ECA). Energy use per unit of GDP is 2.6 times higher than the average for ECA. Industry is the largest consumer of electricity (45 % 2010) and one of the largest sources of energy inefficiency because of the use of outdated and energy-inefficient technology. Agriculture is also one of the most energy intensive sectors of the economy, because of a reliance on an inefficient water-pumping infrastructure.[28] In Tajikistan, the aluminum manufacturing company TALCO accounts for about 40 % of the total net electricity consumption, but does not pay for what it uses.[29]

According to the World Bank report, TALCO could play a crucial role in reducing winter electricity shortages. If TALCO were more energy efficient, household electricity consumption during the winter months could be increased by 25 %.[30]

3. The energy surpluses in the Amu Darya basin countries are often exported rather than being diverted to internal consumption. The main source of electric energy production is gas. In 2010, gas exports were estimated at 14 bcm, which corresponds to a sixfold increase over 2001. Historically, Russia accounted for the largest share of gas exports. In 2002–2010, gas exports to Russia constituted at least 70 % of total gas exports. During the same period, sales to the Kyrgyz Republic and Tajikistan were significantly reduced because of a decrease in demand resulting from price increases and disputes about terms and conditions of gas supply contracts. However, the country has made some progress with diversifying gas exports. Specifically Uzbekneftegaz" and Chinese CNPC will begin the construction of the fourth line of the Uzbek section of the gas pipeline "Central Asia-China" at a total cost of US$800m. The gas pipeline, with a capacity of 20 bcm of gas, is planned to be put into operation in 2017.[31]

In September 2014 the 220 kW and 110 kW networks from Sangtuda-1 and Sangtuda-2 hydroelectric plants transmitted 175.5 million kWh of electricity to Afghanistan, 36.3 million kWh more than in the same period the previous year, and 12.9 % of the total amount of electricity generated in Tajikistan.[32]

Another factor with a decisive influence on the quality and amount of water available in Central Asia, and their possible effects on human well-being, is that the wastes coming from agriculture, energy and human consumption have no appropriate treatment.

AGRICULTURE, ENERGY AND POLLUTION

Agriculture and industry are the main pollutant agents of water in the Amu Darya basin. According to data from the Centre of Hydrometeorological Service, under the Cabinet of Ministers of the Republic of Uzbekistan, published for the Second National Communication of the Republic of Uzbekistan under the United Nations Framework Convention on Climate Change, an increase in water salinity has been observed along the whole watercourse of Amu Darya river. During low water years the mean annual values of water salinity level may increase by 1.5 times, and in certain months by 2–5 times. The data on chemicals in the Amu Darya river water exceed the percentage of maximum permissible concentrations (MPC) rates both upstream (Termez) and downstream (Kyzyldjar). MPC rates are gradually increasing along the river and due to individual ingredients, such as salinity level, magnesium, chlorides, COD.

According to data from the Aquastat survey 2012, quality of both surface water and groundwater is commonly affected by agricultural, industrial and municipal wastewater. In Turkmenistan in recent decades, water quality in the Amu Darya has deteriorated considerably as a result of the discharge of drainage and industrial water from neighboring countries. The average annual salinity level was 0.3 g/liter before 1962, which increased to 0.8 g/liter in 1967. In the 1990s, this stabilized to within the range of 1.5–1.6 g/liter, reaching 2.0 g/liter during certain periods. Human pressure on surface water is high, though pollution caused by biogenic elements or organic substances has not yet reached dangerous levels; special attention must be paid to monitoring concentration (especially of phenols and nitrates). About 4 km^3 of drainage water with salinity level of 6.5–8.5 g/liter is discharged annually into the Amu Darya from neighbouring Uzbekistan.

In Uzbekistan itself, the salinity of irrigation water in the middle reaches of rivers is 1–1.1 g/liter, with a low content of organic substances, and in the lower reaches at certain periods it is, on average, 2 g/liter and more (compared to the initial levels of 0.2–0.3 g/liter), and organic substances 29.6 mg/liter. In some rivers, discharged sewage and municipal wastewater leads to increased pollution all along the course of the river from its origin downstream to the sea. Pollution from petroleum products is 0.4 to 8.2 MAC (maximum allowable concentration), by phenols up to 6 MAC, by nitrates up to 3.7 MAC, and by heavy metals up to 11 MAC. The contamination rate of groundwater has also increased.

In Afghanistan, surface water quality is excellent in the upper basins of all rivers throughout the year and good in the lower basins in spite of the large irrigated areas. Groundwater quality is generally good, but varies from place to place. In the lower reaches of the river valleys, groundwater is frequently saline or brackish and cannot be used for drinking or irrigation. The country faces many environmental problems, mainly the lowering of water tables, the degradation of wetlands, and deforestation (some 40 % of forests have been felled). The excessive use of groundwater for a variety of purposes has significantly depleted aquifers throughout Afghanistan and, if the trend is not reversed, the country will face a severe shortage of drinking water. Recurrent droughts, low precipitation and poor water management have exacerbated the country's water crisis. Over the past several years, groundwater sources have reduced by about 50 %. Limited access to surface water has prompted many farmers, mainly in the drought-stricken south and north, to use groundwater increasingly to irrigate agricultural land, or to dig deep wells. Most of the population uses groundwater as the main, and often the only, source of drinking water.

The lack of adequate treatment of the water from the agriculture, industry and urban sectors is harming the human right to water. Some data on diseases linked to the lack of water quality and a deficient sanitation system make it clear that water quality standards for human consumption in Central Asia are very unsatisfactory. In 2008, according to World Health Organization (WHO) data, mortality rates attributed to the quality of water or deficient sanitation were proportionally higher in Afghanistan, Tajikistan or Uzbekistan than in most Asia-Pacific countries.

In 2015, according to the next UNICEF's map (https://data.unicef.org/topic/child-health/diarrhoeal-disease/), the death rates from diarrhea among children aged under five, a disease linked directly to poor water quality, were highest in Afghanistan and Uzbekistan, similar to those in Central Africa, India and Pakistan.[33]

The poor water quality thus becomes one of the main threats to human security and the human right to water in the Amu Darya region; another of the unresolved threats is the lack of cooperation between the countries of the basin regarding the organization of the uses of water.

In this conflict context, the main cooperation initiatives in the region have been led by the Commonwealth of Independent States (CIS) and the Eurasian Economic Union (EAEU).

THE CIS AND THE EAEU

The CIS failed to generate new economic relations in the area of the former USSR, but provided a good context for new initiatives of economic cooperation in Central Asia. The implementation in January 2015 of the Eurasian Economic Union (EAEU) was the result of a long and complex process of initiatives and treaties of regional cooperation over 25 years. Most of those treaties and initiatives were ultimately fruitless or had a very limited impact, but eventually became the basis for the creation of the current EAEU.

The EAEU Background

The treaties and initiatives of cooperation that finally enabled the creation of the Eurasian Economic Union were the Customs Union formed by the Treaty Establishing the Customs Union between the Russian Federation and the Republic of Belarus (6 January 1995), joined by the Republic of Kazakhstan on 20 January 1995, and by the Kyrgyz Republic on 29 March 1996. On the latter date, the newly-formed "Four" signed the Treaty on Extending Integration in Economic and Humanitarian Spheres that envisaged the creation of a single customs space. On 10 October 2000, the Eurasian Economic Community (EurAsEC) was established by a resolution of the Interstate Council of the Republic of Belarus, the Republic of Kazakhstan, the Kyrgyz Republic, the Russian Federation and the Republic of Tajikistan. In August 2006, at the EurAsEC Interstate Council, a major decision was made to establish a Customs Union between three countries—Belarus, Kazakhstan and Russia. On 6 October 2007, in Dushanbe, capital of Tajikistan, leaders of Kazakhstan, Belarus and Russia signed a treaty on the establishment of a single customs territory, and an action plan for the formation of the customs. In January 2010, the Belarus, Kazakhstan and Russia Customs Union came into effect with the introduction of uniform customs tariffs with exceptions, and a unified customs code came into force. In July 2012 the single economic space was inaugurated and the Eurasian Economic Commission began to function. In January 2015. the EAEU started to be fully operational.

The Eurasian Union Community: Actors and Leadership in Central Asia

During the formation process of the EAEU, its undisputable leader was the Russian Federation, meanwhile Belarus and Kazakhstan played a secondary role; some countries have shown a lower degree of commitment to the aims of the organization, such as Tajikistan and Kyrgyzstan; and some others, such as Uzbekistan, were part of the organization in some of the stages before the current configuration—for example, the case of Turkmenistan, which has stayed out of it from the start. In addition, there are countries initially admitted as observers and eventually as members, such as Armenia in 2015.

Russian leadership within the EAEU has not only been based on Russia's role as a world superpower, its overwhelming size in terms of its economy, population and abundance of natural resources in relation to the rest of the countries, but also in its supremacy in some key bodies of the organization. However, the most powerful body within the EAEU is the Supreme Eurasian Economic Council, where decisions are made by consensus and countries are represented by their most important political representatives in terms of equality. There are some other bodies and mechanisms that favour in particular Russian leadership within the organization:

1. Russia has a great influence on some of the key institutions, such as the Anti-Crisis Fund. This Fund is financed almost entirely by Russia, which secured 88 % of the vote on the Fund's Council Committee. As a decision can only be made once a quorum of 90 % of the votes is obtained, it should not be very difficult for Russia to decide the policy of this new body almost single-handedly. Conversely, nothing can be done without Russia, and it may even be said that the country has a blocking majority.[34]

2. The methods for resolving disputes within EAEU are also favourable to Russia. The basic principle is that of seeking a diplomatic solution to disputes, for which Russia has always had a natural inclination. Consequently, in the event of gross violations of EAEU rules by Russia, such behavior could nevertheless be approved following a diplomatic arrangement. A measure initially contrary to EAEU law could therefore be accepted or imposed on Russia's fellow contracting parties, without the matter coming before the EAEU Court. The flexibility of the EAEU dispute resolution mechanism will therefore undoubtedly allow Russia to impose some of its norms on its partners via the diplomatic route.[35]

Russia thus wants to create a new Eurasia influence area, whose aims, according to Article 4 of the Treaty on the Eurasian Economic Union, are:

(a) To create proper conditions for sustainable economic development of the Member States in order to improve the living standards of their population;

(b) To seek the creation of a common market for goods, services, capital and labour within the Union; and
(c) To ensure comprehensive modernization, cooperation and competitiveness of national economies within the global economy.

In spite of unarguable Russian leadership within the organization, the Central Asian member states, such as Kazakhstan, Kyrgyzstan or Tajikistan, expect that a new economic integration with Russia, based on the values of a free market and freedom of movement of capital, services, resources and people, will improve their economic situation without affecting their political independence. In contrast, Uzbekistan and Turkmenistan consider this organization to be a possible threat against their political and economic independence while the organization suggests a wide frame of cooperation and expects to regulate more than 20 sectors.[36]

EAEU AND WATER RESOURCES

The EAEU has not implemented any initiative, nor has foreseen any cooperation, on water resources in Central Asia, but it has classified the systems of supply and treatment of water as natural monopolies; the result is that each country has a different set of laws and regulations because the organization has not made any progress in regulating the sector, at least at the present time. Of course, the organization's first problem in trying to establish a possible common market for water in the international basins of the Central Asian rivers is that Turkmenistan and Uzbekistan, the largest water consumers in the region, are not members of the EAEU. But it is also important to consider that, to a great extent, Russia still keeps its water policies in the region as only bilateral instead of multilateral.

THE RUSSIAN ROLE, ENERGY AND WATER

Despite the potential that Tajikistan and Kyrgyzstan have to produce hydroelectric power, their energy supplies (gas, oil and even electricity) currently depend on their Kazakh, Uzbek and Turkmen neighbors, because the drought suffered in recent years has exhausted the Tajik and Kyrgyz reservoirs. Current Kyrgyz hydroelectric production is only 10 % of its capacity, and the country suffers under heavy power restrictions. Tajikistan's conditions are not much different from those in Kyrgyzstan, also with continuous limitations on power consumption. Therefore, both countries' energy supply depends on neighboring countries, not only in terms of power, but also of logistics.

The Uzbek territory is the main trade route for Tajikistan and southern Kyrgyzstan. If Uzbekistan border were to close, the Tajik economy would collapse, because the main road and railway hubs, through which the main Tajik exports, aluminum and cotton, are transported, are oriented towards Uzbekistan. Kyrgyzstan is less dependent logistically on Uzbekistan because of its alternative border with Kazakhstan, but the southern area of the country (the Osh province) has similar problems to Tajik territory.

In this context, the Tajik and Kyrgyz governments' freedom of action to achieve their goals is limited; their only chance to carry out their projects and avoid Uzbekistan's pressures as much as possible was to be supported by the traditional region's great power, Russia, which decided to finance the construction of the most important Central Asian hydrologic projects with no restrictions.

However, in January 2009, at a meeting held in Tashkent between the Uzbek president Islam Karimov and Dmitry Medvedev, president of the Russian Federation, Tajik and Kyrgyz hopes of unconditional support for their hydrologic projects seemed to fade. President Medvedev said "the building of hydroelectric stations in Central Asia must take all the states into account and, without a common agreement, Russia would give up those projects". The visit of the Russian to Tashkent ended with a joint Russian—Uzbek statement stating that "the parties have agreed to respect the interests of all the states located on the banks of cross-border Central Asian rivers and the construction of hydroelectric stations will be carried out according to international standards".

The Russian president's statements in Tashkent thus seem to support Turkmen and Uzbek standpoints on the conditions under which hydroelectric complexes must be built, and therefore to give up their old stance apparently supporting Tajik and Kyrgyz arguments.

WHY RUSSIA SEEMS TO HAVE SHIFTED ITS STANCE

The Russian shift of position is probably caused by some energy related questions:

(a) Of Turkmen gas production, 90 % is exported via Russia and seems to be important for the country's fulfillment of its gas contracts, because Russia, according to some specialists, might be in an energy "trough" situation. The Russian capacity for gas exports has apparently reached its limit and the main production fields are in decline. In this situation, if the world demand for gas grows, Russia needs to exploit new fields to increase its production, even though developing this implies large investments and a certain span of time, because a great part of its reserves are located in areas with difficult climatic and orographic conditions (the Arctic coastline and Eastern Siberia). Until the Russian gas reserves can be exploited commercially, Central Asian gas will help Russia to increase its supply offer, or to provide gas for Russia's clients. In that sense, the recent discovery of another very large gas field in Turkmenistan (South Yolatan, with 4–14 trillion cubic meters) and its possible exploitation by Gazpron could have an effect on Russia's stance on water in the area.

(b) As a result of the agreements signed by the presidents of Russia and Uzbekistan at the end of 2009, Russia ensured its commercialization of the Uzbek gas for years to come. The Uzbek president undertook to increase its exports via Russia to 16 billion cubic meters, in addition to the production Lukoil expects to extract from some Uzbek fields that are being explored: some 15–31 billion cubic meters annually. This gas is thought to be transported by increasing the capacity of the current Central Asia Gas Pipeline (CentGas) network passing through Turkmenistan, Uzbekistan, Kazakhstan and the Russian Federation.

Central Asian gas supplements help Russia to strengthen its international position as the world's largest gas exporter, the main export route of Central Asian gas, and at the same time to prevail over possible competitors or alternative routes for gas exports in the region. The gas business might be the reason why the government will not undertake the building of dams, as suggested in the Tajik and Kyrgyz proposals. Russia is thus playing a complex mediator role regarding Central Asian water resources, attempting to satisfy all the actors involved while simultaneously maximizing its own energy-related interests. However, the conflict over the construction of dams between Uzbekistan and its neighbors on the upper course of the river is still simmering in the background.

NOTES

1. *Regional Migration Report: Russia and Central Asia*. Edited by: Anna Di Bartolomeo, Shushanik Makaryan and Agnieszka Weinar. This Report has been published by the European University Institute, Robert Schuman Centre for Advanced Studies, Migration Policy Centre within the frame work of the CARIM-East project. European University Institute 2014. p. 16.
2. Ibid.
3. Ibid.
4. Schneider, Friedrich (2012): *The Shadow Economy and Work in the Shadow: What Do We (Not) Know?* of Labor The Shadow Economy and Work in the Shadow: What Do We (Not) Know? IZA DP No. 6423 March 2012. p. 61. (Forschungsinstitut zur Zukunft der Arbeit Institute for the Study).
5. WIDER Working Paper No. 2010/53 Central Asia after Two Decades of Independence Richard Pomfret. www.wider.unu.edu/ . . . /working-papers/ . . . /wp20 . . .
   OECD Development Centre. Working Paper No 212 (Formerly Technical Paper No. 212). Central Asia since 1991: The Experience of the New Independent States, by Richard Pomfret. www.oecd.org/dataoecd/23/58/5961227.pdf.
6. Civil Society Seminar on Women's Rights. Brussels 21–24 June 2010. Letter of Contract N°2010/234972 Final Report. Prepared by Megan Bick. September 2010 eeas.europa.eu/ . . . rights/ . . . /civil_society/ . . . /2010.EEAS-Europa Central Asia's ratifications of international instruments pertaining to women's rights: http://eeas.europa.eu/_human_rights/dialogues/civil_society/docs/2010_ca_final_report_en.pdf. ILO Sub-regional Office for East Europe & Central Asia: http://www.ilo.ru/gender/cisgnrl.html Civil Society Seminar on Women's Rights. Brussels 21–24 June 2010. Letter of Contract N°2010/234972. Final Report. Prepared by Megan Bick. September 2010 eeas.europa.eu/ . . . rights/ . . . /civil_society/ . . . /2010_ . . .
7. Environment and Security in the Amu Darya Basin. 2011: UNEP, UNDP, UNECE, OSCE, REC, NATO. p. 16.
8. Database Aquastat General Summary for the countries of the Former Soviet Union. http://www.fao.org/nr/water/aquastat/countries_regions/fussr/index5.stm.
9. Peyrouse, Sebastien (2013): *Food Security in Central Asia. A Public Policy Challenge* (Ponars Eurasia Policy) Memo No. 300. September 2013.
10. Khalid, Adeeb (1998): *The Politics of Muslim Cultural Reform: Jadidism in Central Asia*. Berkeley: University of California Press, c1998. http://ark.cdlib.org/ark:/13030/ft8g5008rv/. p. 65.
11. Bakhodyr Muradov and Alisher Ilkhamov (2014): Working paper Open Society Eurasia Program Uzbekistan's Cotton Sector: Financial Flows and

Distribution of Resources (Matt Fischer-Daly and Jeff Goldstein, Editors). October 2014. http://www.opensocietyfoundations.org/sites/default/files/uzbekistans-cotton-sector-20141021.pdf.

12. Ibid.
13. Ibid.
14. USDA October 2012. Report from the Economic Research Service Economic Policy and Cotton in Uzbekistan. Stephen MacDonald, Economic Policy and Cotton in Uzbekistan/CWS-12 h-01. p. 3.
15. Ibid., p. 4.
16. Bakhodyr Muradov and Alisher Ilkhamov (2014): Working Paper Open Society Eurasia Program Uzbekistan's Cotton Sector: Financial Flows and Distribution of Resources (Matt Fischer-Daly and Jeff Goldstein, Editors). October 2014. p. 11. http://www.opensocietyfoundations.org/sites/default/files/uzbekistans-cotton-sector-20141021.pdf.
17. Ibid.
18. Ibid., pp. 12–13.
19. Yulia Shirokova and Alexander Morozov, "About Ways for Improvement of Water Use in Irrigation of Uzbekistan", pp. 358, 363.
20. Soil Quality Resource Concerns: Salinization", U.S. Department of Agriculture, January 1998. "Land Resources—Uzbekistan", FAO, www.fao.org/nr/land/projects programmes/cacilminitiative/cacilm-project/uzbekistan/e.
21. International Crisis Group Europe and Central Asia Report No.233 11 September 2014 p.
22. EIA. http://www.eia.gov/countries/country-data.cfm?fips=UZ Ids (Galkynysh) boosted its economy in 2013.
23. Ibid., Tajikistan has an installed capacity to produce about 527 billion kWh a year, second among CIS Countries, eighth in the World. This is three times more than the present annual consumption of Central Asia. http://www.unece.org/fileadmin/DAM/commission/2011/Presentation_speeches/2ndDay/01_Hamrokhon_Zarifi.pdf.
24. ABD Poverty and Social Analysis and Strategy. http://www.adb.org/sites/default/files/linked-documents/45120-003-sprss.pdf. Uzbekistan 45120-003 Samarkand Solar Power Project.Approved 20 November 2013.
25. German Development Cooperation with Afghanistan Expansion of electricity supply. http://www.ez-afghanistan.de/fileadmin/content/factsheets/englisch/2012-01-energie-sektor-e.pdf.
26. Oskarsson, Katerina (2012): *Energy Development Security Nexus in Afghanistan*, 20 November 2012. http://ensec.org/index.php?option=com_content&view=article&id=386:energy-development-security-nexus-in-afghanistan&catid=130:issue-content&Itemid=405.
27. Ibid.

28. 2013 International Bank for Reconstruction and Development/The World Bank. Uzbekistan Energy/Power Sector Issues. Note p. 17.
29. "TALCO Energy Audit: Improved Efficiency Could Help Solve Winter Electricity Shortages", Fact Sheet, World Bank, December 2012; "Financial Assessment of Barki Tojik", World Bank, October 2013, p. 6.
30. TALCO Energy Audit: Fact Sheet. December 2012. Improved Efficiency Could Help Solve Winter Electricity Shortages. http://www.worldbank.org/content/dam/Worldbank/document/tj-talco-energy-audit-fact-sheet.pdf.
31. 2013 International Bank for Reconstruction and Development/The World Bank. Uzbekistan Energy/Power Sector Issues. p. 6.
32. Monitoring and Early Warning in Tajikistan Monthly Report. October 2014, Ministry of Economic Development and Trade. http://reliefweb.int/sites/reliefweb.int/files/resources/TJK_Monitoring_and_Early_Warning_Report_October_2014_ENG.pdf. Tajikistan 37 Bokhtar, 2nd floor, office #225, Dushanbe 734002.
33. Levels & Trends in Child Mortality Report 2013. Estimates Developed by the UN Inter-agency Group for Child Mortality Estimation United. UNICEF http://www.childinfo.org/files/Child_Mortality_Report_2013.pdf. p. 11.
34. Russian Law Journal Volume III (2015) Issue 1.12, Russia's Normative Influence over Post-Soviet States: The Examples of Belarus and Ukraine, Hugo Flavier. Montesquieu University—Bordeaux IV (Bordeaux, France).
35. Ibid., p. 28.
36. Customs tariff and non-tariff regulation; customs regulations; technical regulations; sanitary, veterinary-sanitary and phytosanitary quarantine measures; transfer and distribution of import customs duties; establishment of trade regimes for third parties; statistics of foreign and mutual trade; macroeconomic policy; competition policy; industrial and agricultural subsidies; energy policy; natural monopolies; state and/or municipal procurement; mutual trade in services and investments; transport and transportation; monetary policy; intellectual property; labour migration; financial markets (banking, insurance, the currency market, the securities market); other spheres as specified in the Treaty and other international treaties within the Union related to the economy; however, none of them makes any explicit reference to water.

CHAPTER 5

Political Security Water Resources and International Security Organizations (NATO, OSCE, CSTO, SCO) in the Amu Darya

Abstract To Josef Nef, political security means the guarantee of several rights, such as the right to be represented, to participate, and to disagree, but in particular to have a real chance to make effective changes in the laws and regulations of social and political life. Therefore, protecting, respecting and developing the Human Right to Water and Sanitation (HRWS), as in the case of any other human right, not only involves preserving the amount and quality of water resources, but also ensuring the participation of all users (citizens, communities, non-governmental organizations (NGOs), economic sectors, etc.) in the economic and emancipation possibilities involved in the management of water as a resource. The achievement of the HRWS is thus a political challenge and not merely a simple distribution of the resource that enables individuals or groups to participate in the decision-making process freely and unthreatened. In this chapter, the author analyzes the connection between political systems, borders, ethnic groups and water resources, and how they affect the human right to water.

Keywords Security · Human rights · Corruption · Sultanism · Ethnic conflict · Multilateral security organizations · Terrorism

M.Á. Pérez Martín, *Security and Human Right to Water in Central Asia*, Security, Development and Human Rights in East Asia, DOI 10.1057/978-1-137-54005-8_5

INTRODUCTION

Josef Nef (1999) define the political security as the right to representation, autonomy (freedom), participation and dissent, combined with empowerment to make choices with a reasonable probability of effecting change.

To Nef, political security means the guaranteeing of several rights, such as the right to be represented, to participate and to disagree, but in particular to have a real chance to make effective changes in the laws and regulations of social and political life.

Therefore, protecting, respecting and developing the HRWS, as in the case of any other human right, not only involves preserving the amount and quality of water resources, but also to ensure the participation of all users (citizens, communities, NGOs, economic sectors, etc.) in the economic and emancipation possibilities involved in the management of water as a resource. In the words of Ken Booth, "'security' means the absence of threats," but also "of those physical and human constraints which stop them carrying out what they would freely choose to do" (Booth 1991, p. 319). Political, social or cultural oppression are direct threats to the freedom of individuals, human rights and the human right to water.

The achievement of HRWS is thus a political challenge and not just a simple distribution of the resource that enables individuals or groups to participate in the decision-making process freely and unthreatened. Consequently, the nature of the political systems involved will be a key element in the achievement of the HRWS goals, because these are the contexts where the states must implement this right. In this chapter we shall analyze the connection between political systems, borders, ethnic groups and water resources, and how they affect the human right to water. And we shall also examine the policies of international organizations dealing with conflicts over water resources.

HRWS IN CENTRAL ASIA

The percentages of the Central Asian population's access to water and sanitation, shortly before its recognition as a human right by the United Nations in July 2010, were as shown in the next table (Table 5.1).

This table shows high percentages of the population having access to water and sanitation in Central Asian cities, but they are much lower in rural areas. We shall now analyze accessibility to water and the right to sanitation of in the rural environment, where most of the region's population lives.

Table 5.1   Accessibility to water and sanitation

	Accessibility to water % population		With sanitation % population	
	Urban	Rural	Urban	Rural
Kazakhstan	78	35	84	10
Kyrgyzstan	82	58	68	28
Tajikistan	93	49	20	5
Turkmenistan	85.4	42.1	61.8	2
Uzbekistan	90	71	85	40

Source: Personally elaborated with data from Review 2009. Water GEF Fond Environment Global. Regional Review Water Supply and Sanitation in the Countries of Central Asia and Southern Caucasus. WSSOzor_ENG_Aug2009.pdf http://www.gwp.org.

Table 5.2   Rural population percentages

	Total population 2011	Rural population percentage (%)	Total rural population
Uzbekistan	29,341,200	63	18,484,956
Kazakhstan	16,558,459	42	6,954,553
Turkmenistan	5,105,301	51	2,603,704
Tajikistan	6,976,958	74	5,162,949
Kyrgyzstan	5,507,000	63	3,469,410
Total	63,488,918	59	37,204,506

Source: Personally elaborated with data from Review 2009. Water GEF Fond Environment Global. Regional Review Water Supply and Sanitation in the Countries of Central Asia and Southern Caucasus. WSSOzor_ENG_Aug2009.pdf. http://www.gwp.org.

As the next table shows, most of the people in Central Asia—almost 59 %—live in rural areas: more than 37 million people. The countries with the largest proportions of rural population are Tajikistan (74 %), Kyrgyzstan (63 %), Uzbekistan (63 %), Turkmenistan (51 %) and Kazakhstan (42 %) (Table 5.2).

We shall now examine the figures and percentages of the rural population with no access to water in each country (Table 5.3).

The countries with the lowest percentages of access to drinking water are led by Turkmenistan—68 % of its population has no access to potable water; followed by Kazakhstan (65 %), Kyrgyzstan (52 %), Tajikistan (51 %) and Uzbekistan, the country with the best results in this case, as only 29 % of its population has no access to water (though in absolute terms this means more than 5 million people). There are more than 19 million people living in the region without access to drinking water—53 % of the whole rural population.

Table 5.3   Percentage of rural population without access to water

	Rural population	*Percentage of inaccessibility to water (%)*	*Total population without accessibility to water*
Uzbekistan	18,484,956	29	5,360,637
Kazakhstan	6,954,553	65	4,520,459
Turkmenistan	2,603,704	68	1,770,519
Tajikistan	5,162,949	51	2,633,104
Kyrgyzstan	3,469,410	52	1,804,093
Total	36,675,572	53	19,438,053

Source: Personally elaborated with data from Review 2009. Water GEF Fond Environment Global. Regional Review Water Supply and Sanitation in the Countries of Central Asia and Southern Caucasus. WSSOzor_ENG_Aug2009.pdf http://www.gwp.org.

Table 5.4   Percentage of rural population without sanitation

	Rural population	*Percentage without sanitation (%)*	*Total population without sanitation*
Uzbekistan	18,484,956	60	11,090,974
Kazakhstan	6,954,553	90	6,259,098
Turkmenistan	2,603,704	98	2,551,630
Tajikistan	5,162,949	95	4,904,802
Kyrgyzstan	3,469,410	70	2,428,587
Total	36,675,572	83	30,294,022

Source: Personally elaborated with data from Review 2009. Water GEF Fond Environment Global. Regional Review Water Supply and Sanitation in the Countries of Central Asia and Southern Caucasus. WSSOzor_ENG_Aug2009.pdf http://www.gwp.org.

With regard to the population without the right to sanitation, the figures are as in the following table (Table 5.4).

Turkmenistan has the highest percentage of the rural population with no right to sanitation, 98 %; second is Tajikistan (95 %); third, Kazakhstan (90 %); fourth, Kyrgyzstan (70 %); and finally Uzbekistan (60 %). However, as in the case of access to water in absolute terms, the latter is the country with the largest number of people without the right to sanitation: more than 11 million. More than 30 million people in Central Asia have no guarantee of their right to sanitation: 83 % of the whole rural population.

The most recent information about the progress being made with regard to drinking water and sanitation published by the World Health

Organization (WHO) in 2014 warned that the Caucasus and Central Asia region are the only ones in the world that, according to the Millennium Development Goals (MDG), recorded a slight decline in drinking water coverage: around 1 % less of its population.[1]

Approximately 7.5 million of the 28.9 million people in Uzbekistan, and 4.8 million of the 8.05 million in Tajikistan, lack adequate access to clean drinking water. Roughly 2 million of Kyrgyzstan's 5.6 million also lack such access. WHO notes there had been some growth in "improved access to water" in Central Asia after 1990. But debate exists among water experts about what qualifies this as an improvement—it could mean as little as one public tap serving an entire village. The Swiss Agency for Development and Cooperation (SDC), which began working in the three countries in the early 1990s, observed that since the collapse of the Soviet Union, "fewer and fewer people have access to clean water because the budgets of the newly independent states contain very limited funds to build new water infrastructures for the rapidly growing population. Existing systems fall into disrepair or break down altogether because no funds are available to maintain them".[2]

Human rights violations related to water resources in the Amu Darya basin are not determined only by the decisions, or the lack of them, taken by the states with regard to the use, access to or the quality of the water, but also by the building of large dams and water infrastructure works, which can threaten the human security of millions. According to a Human Rights Study (2014), the Rogun dam construction in Tajikistan will cause the displacement of more than 42,000 people. Between 2009 and early 2014, the government had already resettled approximately 1,500 families from the reservoir zone to several other locations in Tajikistan. Based on interviews with people at various stages of the resettlement process, Human Rights Watch has found that the standard of living for many resettled families has seriously deteriorated, and that there are a number of barriers undermining their ability to re-establish the standard of living they enjoyed prior to being resettled. Loss of land for farming and raising livestock, lack of employment, and poor access to essential services in resettled communities have combined to create significant hardship for resettled families, seriously diminishing the exercise and enjoyment of their fundamental rights. On 17 June 2014, the World Bank published the final draft of its Rogun dam studies for consultation, as well as its own draft paper, "Key Issues for Consideration on the Proposed Rogun Hydropower Project".

The World Bank acknowledged that the required resettlements would have a major impact on building the Rogun dam, that the project would result in economic, as well as physical, displacement, and that restoring livelihoods during and after resettlement would be a critical element of the resettlement process. However, while the draft Environmental and Social Impact Assessment (Garcés de los Fayos 2014) importantly considers international environmental treaties and international water laws, it does not consider relevant international human rights instruments regarding resettlement.[3]

Despite the catastrophic condition of the HRWS in the region, in July 2010 only two countries decided to recognize access to water and sanitation as a human right: Tajikistan and Kyrgyzstan. Kazakhstan abstained and Uzbekistan and Turkmenistan were not present for this vote.

Those countries with an abundance of water and water resources (Kyrgyzstan and Tajikistan) are more willing to recognize, at least on paper, the human right to water, while those consuming larger amounts of water and without the possibility of the renovation of their water resources (Kazakhstan, Turkmenistan and Uzbekistan) are less prepared to do it. However, there exist other reasons, of a political nature, that also affect this lack of recognition of the human right to water. In this sense, it is especially noticeable that countries such as Uzbekistan, Turkmenistan and southern Kazakhstan, which depend on their neighbors' water resources, do not recognize the human right to water, with the purpose of demanding from both Tajikistan and Kyrgyzstan a guarantee already recognized by them.

These facts lead one to think that the allocation of water resources in Central Asian countries is carried out arbitrarily, without much transparency and in an authoritarian way. There is an illuminating article on this question, by Kai Wegerich: "*Natural Drought or Human Made Scarcity in Uzbekistan?*" (Wegerich 2002). The author's conclusion is that some droughts in the lower course of the Amu Darya are not caused by natural phenomena or technical and institutional problems regarding water management, but by the Uzbek authorities' decision to increase cotton production. This also discourages interstate cooperation, as this would involve more transparency in terms of information exchange about the Amu Darya's volume of water in its course through the different states and, as a consequence, about the transfer of water to undeclared activities. The water crisis is therefore not based on absolute physical limits but rather based on historical lines of resource capture (of both land and water) and ecological marginalization.[1]

Political decisions about by whom, how much, when and how water resources are managed in the region are determined mainly by values and practices that are not transparent. The implementation of the human right to water and sanitation (HRWS) might be an important tool to replace these practices and to promote individual empowerment, as well as to guarantee human rights for many more people in Central Asia.

WATER RESOURCES AND ETHNIC CONFLICTS

Ethnic tensions in Central Asia usually have a socioeconomic background, often related to the control and supply of the main economic engine of the region: water resources. These kinds of water conflicts can strengthen and trigger disagreements, mainly between ethnic groups living near the higher courses of the main Central Asian river basins and those settled in the medium or lower courses; or between ethnic groups with infrastructures to control water and those who lack them.

Professor Stephan Klotzil has documented a series of conflicts between ethnic groups with the background of water resources control or distribution in Central Asian hydrographic basins. The region's conflicts listed in the next table can be of different kinds (intrastate or interstate) and connected with water resource distribution or border delimitations affecting the ownership and management of hydrologic infrastructures (reserves, dams, canals) or the distribution of lands with irrigation infrastructure (Table 5.5).

The dimensions of the conflict in this watercourse also affect the states (Kranz et al. 2005), and all who are connected with those infrastructures and projects related to water resource management and use of the river:

1. Tensions between Tajikistan, Uzbekistan and Turkmenistan because of the construction of the new Rogun dam in Tajik territory.
2. Tensions between Turkmenistan and Uzbekistan over the management of the Karakum canal.
3. Tensions between Turkmenistan and Uzbekistan because of the management of the Tujamujun dam.
4. Tensions between Turkmenistan and Uzbekistan over the construction of "The Golden Age" dam in Turkmen territory.

Table 5.5   Conflicts related to water in the Aral Sea basin

Hydrologic systems	Control of supplies	Main users	Intrastate conflicts or ethnic-territorial conflicts	Degree of conflict seriousness
Naryn and Toktogul Reservoir	Kyrgyzstan	Kyrgyzstan Uzbekistan	Ethnic tensions between Uzbeks and Kyrgyzs in the Fergana Valley	High
Karakorum Reservoir	Tajikistan	Uzbekistan Tajikistan	Water transfer from the Tajik section of the Fergana Valley to Uzbekistan	Medium
Tributaries to the Fergana Valley	Kyrgyzstan	Uzbekistan Tajikistan	Ethnic tensions between Uzbeks and Tajiks	High
Chardara Reservoir	Kazakhstan	Kazakhstan Uzbek minority	Land transfers between the Syr Darya and the Arys (Chimket province) from Kazakhstan to Uzbekistan	Low
Vakhsh/ Pyandsh	Tajikistan	Tajikistan	Factions distributed along the Amu Darya course between Gorno Badakhstan and the region of Kurgan Tyube	High
Zeravshan	Tajikistan	Uzbekistan	Ethnic tensions between Uzbeks and Tajiks; transfer of water resources from Zeravshan River to other places	Medium
Lower Amu Darya	Turkmenistan and Uzbekistan	Turkmenistan and Uzbekistan	Territorial claims related to water reserves; Tazhaus Oasis, provinces of Khorezm and Cardzhou in the Amu Darya middle basin	Medium

Source: A Source of Conflict or Cooperation in Central Asia PLM 5620—Research Project (Politics), p. 49, Appendix 2c. Table 2: Water-related conflicts in the Aral Sea basin (Klotzli, 1994: 43). Allouche, J. (2004). A source of regional tension in Central Asia: The case of water. The Illusions of Transition: which perspectives for Central Asia and the Caucasus?, Geneva, Switzerland, CIMERA Graduate Institute of International Studies (IUHEI) Vol. 6.

We shall now analyze the main water conflicts in the region, and the cause of all the current conflicts related to water: the Aral Sea desiccation.

The Aral Sea desiccation

During the Soviet era, the central government in Moscow controlled the entire network of rivers shared among its republics through water-use quotas. This approach meant that the borders between the Central Asian republics had little, if any, effect on basin management. The collapse of the Soviet Union and the creation of new states fragmented water management in the region. With the aim of avoiding conflicts in this new political context, Central Asian countries formed the Interstate Commission for Water Coordination (ICWC), established in accordance with the "Agreement on collaboration in the sphere of joint water resources management within interstate water sources" dated 18 February 1992, and approved by the heads of state on 23 March 1993. The ICWC is a collective body that manages transboundary river courses and is responsible for water allocation among countries; monitoring; and preparing preliminary assessments of proposals on institutional, ecological, technical and financial approaches, based on decisions mutually agreed by all sides. The two Basin Water Organizations (BWOs) (Amu Darya and Syr Darya), the Scientific-Information Center, and the ICWC Secretariat are executive bodies of this Commission.

The ICWC took over responsibility for water management in both basins directly from the former Soviet Ministry of Water Resources, but with appropriate changes reflecting the creation of five new independent states:

- The Commission has five members, appointed by the governments. They are equal in rights and obligations. They meet once a quarter to decide on all issues related to their activities and responsibilities. Decisions are reached only on a consensus basis.
- Two BWOs were transformed into the executive bodies of the ICWC; in a similar way, a part of the Central Asian Scientific Institute for Irrigation (SANIIRI) was transformed into the Scientific-Information Center (SIC) of the ICWC to act as a think tank for the commission.
- All issues for the ICWC meetings, in accordance with their agenda, are prepared by the executive bodies and disseminated among the members 20 days before each meeting; this allows for the preparation of comments and opinions by each country.

• The principles of water allocation that existed in Soviet times have been retained for the purpose of annual planning until new regional and national water management strategies can be developed and adopted.

According to the Decision by the Heads of State of 23 March 1993, the ICWC was included in the International Fund for saving the Aral Sea (IFAS) and has the status of an international organization, and later by an agreement among the region's five countries, on 9 April 1999: "On the status of IFAS and its organizations".

THE INTERNATIONAL FUND FOR SAVING THE ARAL SEA

The International Fund for Saving the Aral Sea (IFAS) is an international organization created in 1993 with the aim of recovering the Aral Sea's environment. It has five Central Asian countries as members, as well as a several donors: the European Commission, the European Parliament, the European Union (2010), the Organization for Security and Co-operation in Europe (OSCE), the World Bank, the Central Asia Regional Economic Cooperation (CAREC), the United States Agency for International Development (USAID), the German Corporation for International Cooperation (GIZ) (2010), the United Nations Economic Commission for Europe (UNECE) (2010), the Eurasian Development Bank (EDB), the Swiss Agency for Development and Cooperation, the United Nations Regional Centre for Preventive Diplomacy for Central Asia (UNRCCA) (May 2010), and the International Office for Water (IOWater).

The mission of this Fund is to coordinate cooperation issues in two spheres (national and international), with the aim of using water resources in the most effective way and to improve the Aral Sea's socioeconomic conditions. The Fund thus acts as a platform for dialogue between Central Asian countries and international donors.

Since its creation, the IFAS has developed a number of action programs (Aral basin program-1, Aral basin program-2), with each program having different aims. The first program (1993–1994) had the following aims:

1. Stabilize the environment in the Aral basin.
2. Recover the environment in the Aral Sea coastline area.
3. Improve the cross-border management of water resources.
4. Develop the regional organizations' capacity to carry out the plan.

These were the purposes of the second program (2002):

1. Carry out a series of projects to improve the environment.
2. Socioeconomic conditions.
3. Improve the institutions involved in cooperation.

These two programs, which ended in 2010, were not very fruitful, especially in terms of cross-border cooperation regarding water resource management. In one of its reports, the Executive Committee of the IFAS established some of the reasons why the programs have not been very successful (Serving the People of Central Asia 2011):

> Up to now, the main purpose of our projects has been related to technical questions, paying little attention to the social, political and institutional ones. There has not been enough cooperation between ministries of the involved countries, local authorities and civil society, because these actors have not always been part of the decision-making process. As a consequence, these programs have not sufficiently attracted the public's attention.

However, the Committee also acknowledged some positive effects:

> However, the programs have substantially helped to link the region's countries and the international donors. This fact has strengthened the countries' capacity to automatize strategies and to establish priorities for the economic development of the social sector and water resources management.

The Executive Committee's report basically recognizes the great complexity of water resource management, because it does not only involve technical questions, but also plenty of economic, political and social factors with more than single actors in them, at different levels and with different responsibilities, and all of them are necessary in order to make a decision.

After their previous experience, the IFAS has focused its efforts on a third program, whose final aim is:

> The improvement of the socioeconomic and environmental conditions by applying the integrated water resources management principles, developing acceptable mechanisms for a varied use and protecting the Central Asian environment, taking into consideration the interests of all the states in the region.

And the works to be developed will have four different aspects:

1. Integrated management of water resources. Including projects related to the implementation of monitoring systems, databases and security systems for infrastructures linked to water. The aim is creating an information network and monitoring indexes equivalent in all the five countries of the Aral basin.
2. Environmental protection. Including projects related to the preservation of biodiversity and plans of environmental risks reduction. The purpose is improving the current environmental conditions in the region as well as preventing possible new threats.
3. Socioeconomic development. Includes measures related to the creation of jobs, and the implementation of sustainable development policies, education and public health, with the aim of strengthening human resources.
4. Strengthen institutions and legal instruments. Includes proposals for institutional and cooperative development in environmental terms, as well as strengthening the roles of some regional actors, such as ministries and civil society, with the aim of improving cross-border cooperation.

In brief, the cross-border cooperative experience with water resources is still in its early stages in the region, though some of the projects have begun to be fruitful. For example, in the Kazakh area of the Aral Sea, because of a dam built in collaboration with the World Bank, water accumulation increased the dam level from 30 meters in 2003 to 42 meters in 2008, though the recovery is not yet complete, because it is a very fragile ecosystem. The sea is productive again and commercial fishing is being carried out once more; but this recovery process has not yet begun in the Uzbek zone within the area of the Aral Sea, and exploratory drillings are being undertaken by several oil companies to find either gas or oil.

These conditions of environmental and human catastrophe are not comparable to the rest of the Amu Darya basin, but the situation regarding the quality of water and its safety is of great concern, as we have discussed in previous chapters. Despite the political and socioeconomic changes in the region after the dissolution of the USSR, and the inclusion of the former Soviet republics as independent states among the international community, the environmental risks continue to threaten extensive areas in Central Asia, and the degradation of the quantity and quality of the resources in the Amu Darya basin is evident.

This situation has been caused mainly by as range of political factors, such as authoritarian political systems, regional rivalries, ethnic conflicts and the failure of international security organizations in their defense of human rights.

POLITICAL SYSTEMS

More than two decades have passed since the republics existing now in Central Asia won their independence. One of the main concerns for the scientific community was about the political path these new states would take after the collapse of the USSR; would they become democratic states, Islamic republics or autocracies difficult to be defined? Today, more than 20 years later, one of the key questions is still the same: are the political systems of the Central Asian republics evolving to become states of law and democracies? After decades of study and reflection, something is obvious in Central Asia: Central Asian states are not democracies; what, then, are they?

All those with specialist knowledge about the region agree on classifying Central Asian countries as hybrid, neo-patrimonial or authoritarian states to a greater or lesser degree. One of the first jobs that established a political typology for the new Central Asian regimes and their differences is the compilation of articles edited by Sally N. Cummings, titled "*Power and change in Central Asia*" (2002). Central Asian political systems are labeled in this compilation as follows: Muriel Atkin defines the Tajik political system as "ineffective authoritarianism;" the Uzbek system is defined by Roger D. Kangas as "benign authoritarianism;" the Kyrgyz system is called by Eugene Huskey, "minimalist authoritarianism;" and, finally, Kazakhstan is labeled by Sally N. Cumming as a "hybrid regime in transition". There is another classification by Oleksandr Fisun (2003, p. 5–6),[5] who states that Turkmenistan and Uzbekistan are sultanistic and neo-patrimonial semi-competitive regimes; Tajikistan and Kyrgyzstan are neo-patrimonial oligarchies; Kyrgyz was also semi-competitive during Askar Akayev's government between 1990 and 2005; and in the Tajik case there is low competitiveness. These categorizations of Central Asian political regimes are influenced, to a greater or lesser extent, by the contributions of Berg-Schlosser (2007), who coined the term "sultanism." This concept has had a great impact on Central Asian political studies. Sultanism is defined by Berg-Schlosser as "a regime in which all the individuals, groups and institutions are permanently subject to the unpredictable and despotic intervention of the sultan and, as a consequence, all pluralism is precarious". These are the main features of sultanism[6]:

(a) Pluralism; political, economic and social pluralism does not disappear, but becomes the object of unpredictable and despotic interventionist policies. No group or individual is able to escape the despotic power of the sultan. The rule of law does not exist. Low degree of institutionalization. High degree of fusion between the public and private spheres.

(b) Ideology; the symbols suffer a strong and arbitrary manipulation. Extreme glorification of the leader. There are neither elaborated ideology guidelines nor any distinctive mindset, apart from despotic and personal characters. There are no attempts to justify initiatives on an ideological basis.

(c) Mobilization; there is a low degree of mobilization, occasionally manipulated with ceremonial aims either by coercion or political favoritism. Periodic mobilization of state-sponsored groups using violence against other groups encouraged by the sultan.

(d) Leadership; personal and arbitrary. Neither rational-legal nor ideological restrictions. Acquiescence towards the leader based on strong fear and personal rewards. Recruitment of personnel is from among family friends, business partners or individuals involved in the use of violence to support the regime. Stances and attitudes are determined exclusively by personal submission to the ruler.

In this political context, water management is defined by a vertical and strongly hierarchized organization, where all the decisions are made by central governments with no participation by local actors (Wiegmann 2011, p. 5). This type of organization caused some problems, among them the fact that guidelines from the ministries very often do not match local needs; this, and the low salaries of public servants, has aided corruption and the use of influence to come into play, as happened during the Soviet era, which has been an important factor in the allocation of water resources (Micklin 2000, p. 64). On the other hand, the participation of NGOs, or associations of users and professionals in matters of drinking water supply, sanitation and environment, is usually understood in a very restricted sense, like the right of these actors to be informed about water management but not to participate in the decision-making processes. The Global Environment Fund (GEF) has carried out a survey in an attempt to analyze the degree of public

Table 5.6 Degree of public participation in each country of the region

	Availability of laws encouraging public participation	Implementation degree of laws	Experts' opinion about public intervention
Kazakhstan	Available	Normal	Medium
Kyrgyzstan	Available	Initial	Low
Tajikistan	Available	Low	Unsatisfactory
Turkmenistan	None	None	None
Uzbekistan	None	None	None

Source: Carried out with data from Review 2009. Water GEF Fond Environment Global.

participation in the management of water resources. The outcome is shown in the table (Table 5.6).

According to these data, Uzbekistan and Turkmenistan are less enthusiastic regarding public participation in their water resources; Tajikistan has some legislation on this matter, but it is not effective; Kazakhstan has a partial degree of public participation; and Kyrgyzstan is just at the beginning. The lack of transparency and equity in the allocation of water resources, as well as water distribution based on networks of political favoritism and vested interests, is seriously harmful to the government's legitimacy and the consideration of water as a human right, because water resources are managed only as an economic good on the fringes of the law.

HRWS AND WATER: ETHNIC CONFLICTS IN THE AMU DARYA WITH THE CONSTRUCTION OF THE NEW POST-SOVIET STATES

Throughout the 70 years of Soviet dictatorship, communist leaders proclaimed that the universality of the "Soviet man" and the "Soviet system" would overcome the great diversity and many cultural peculiarities of the Russian Empire's peoples that composed the Soviet Union at that time.[7]

At the beginning of the 1920s, Moscow's Soviet started a new territorial political agenda that to some extent contradicted the Marxist principles of homogenization and universality previously proclaimed, marking out new interior borders and creating national entities within the Soviet space. The national delimitation of borders in Central Asia was carried out in different stages, the most important one being between 1924 and 1936. Most of the borders drawn in this region obeyed a political agenda based on the motto *"Divide et impera"*, necessary for Moscow's political management and control of this diverse and huge area, by means of borders lacking any historic, geographic, economic or ethnic rationality.[8]

The classification of its inhabitants in terms of nationality was determined by criteria developed by the Soviet authorities, and Stalin, in turn inspired by the nationalistic ideas of the nineteenth-century German Romantic School, had a large influence on Russian ethnography.

The general principles of ethnic or national classification were:[9]

1. Each ethnic group has a territory, and the ascription of individuals to that ethnic group is determined by their mother tongue.
2. A political-administrative system based on the "territorialization" of language.
3. A real practice of peoples' and territories' delimitation not based exclusively on the principles previously mentioned, but also with a strategic and political logic, though expressed in terms similar to those already described.

Therefore, in theory, a language constituted a nation with a guaranteed territory and a specific administrative status; in fact, major ethnic groups in Central Asia were considered, though not always, to be "title holders" in their correspondent state. The political practice of national delimitation gave way to several incongruences:

1. Major ethnic groups living in the same administrative territorial frame, but sharing their space with other minorities, which in turn are the major groups in neighboring states. This happens in all the region's countries, because all have important minorities in every bordering country; the most remarkable case is Tajiks in Uzbekistan, or Russians in Kazakhstan.
2. Some ethnic groups who have no native tongue, but with a local or regional cultural tradition different from communities with the same language, were denied access to territorial claims: Turks, Meskhetians, Tatars, etc.
3. Ethnic groups with a native tongue and culture whose political, economic or demographic relevance was small compared to other groups; they were not offered any special territorial or administrative status but, on the contrary, were given a degree of autonomy

(oblast, autonomous republic) subject to a higher administrative frame, the Soviet Socialist Republics, ruled by a different ethnic group: Karakalpaks, Pamiris, etc.

The experiment of devising new nationalities assigned to certain political-administrative units (states) was complemented, a few years later, by the arrival in the region of new nationalities and ethnic groups, the so called "punished peoples", as well as a new wave of Russian settlers.

THE PUNISHED PEOPLES

Throughout the period 1937–1945 more than 3 million people were forcibly deported to the republics of Central Asia and Siberia according to two different criteria:[10]

1. Those who had migrated to Russia from countries by then at war with the Soviet Union, such as Korea, Romania, Germany, Finland, Greece; and Meskhetian Turks and Kurds.
2. People belonging to ethnic groups with historic claims of independence or autonomy from Moscow: Karachays, Chechens, Ingush, Balkars and Crimean Tatars.

Both groups were considered to be a danger to the national security during the World War II. Stalin feared they might create a fifth column to support a Nazi invasion. They were thus confined in labor camps thousands of kilometers away from their homes, in inhospitable places, and they were underfed. Hundreds of thousands died during the deportation and settling process in the new labor camps.

Some of the preferential destinations for the deportees were the republics of Kazakhstan, Kyrgyzstan and Uzbekistan, at the present time housing important minorities, especially Koreans, Tatars and ethnic groups from the Caucasus. These nationalities or ethnic groups coexisted during the Soviet era with the Central Asian native populations, always, of course, subject to Moscow's authority; but after the Central Asian republics' independence, some of the ethnic groups originally deported, such as the Meshketian Turks of Uzbekistan, were the focus of persecutions leading to a new exodus towards, in this case, their old home in Georgia.[11]

Coexistence among major ethnic groups, or "title holders", and minor ones, or "non-title holders", is often difficult and tense, and has become one of the most relevant challenges and threats for territorial stability and integrity in the new Central Asian states.

THE CONSTRUCTION OF THE NEW POST-SOVIET STATES. COMMUNITIES AND BORDERS

When the republics proclaimed their independence, a series of processes of national construction began, forcing the inhabitants of every state to declare themselves "Uzbeks," "Turkmens," "Tajiks" or "Kyrgyz" (except in Kazakhstan) independently of their nationality; that is, to declare themselves members of the nationality which, considered as the "title holder" during the Soviet era, ruled the destiny of the new independent republics. The Central Asian states, to a greater or a lesser extent, have started national construction processes basing their identities on the "ethos" of the "title-holder" ethnic group in the "Uzbekization," "Turkmenization," etc. of the society, together with political agendas using discriminatory language, and a political and cultural marginalization of their minorities, or even the criminalization of these in some cases.[12]

LEGAL FRAMEWORK AND MINORITIES

All the constitutions of Central Asian countries recognize the equality of the rights and freedoms of their citizens, independently of their race, religion, ethnic group, etc. All have signed the main international conventions involving commitment to respect human rights, and some of them, such as Uzbekistan, have even set up posts and bodies such as the Commissioner of the Parliament for Human Rights, the National Center for Human Rights, or the Institute for Civic and Social Studies. These have only a consultative character, with no capacity to bring possible violations of human rights to court. But the practical application of these conventions, or even national laws, and the working of bodies related to minorities' rights, are very weak or simply do not exist.

Respect for language and racial diversity is also a right recognized by all the constitutions of the Central Asian countries. In some cases there are languages considered equally official, such as Kyrgyz (Kyrgyz-Russian) or Tajik (Tajik-Russian), but language minorities always blame their governments for not assigning the necessary resources to implement, in schools, high schools and universities, the languages that do not belong to the "title-holder" ethnic group in each republic. Meanwhile, most of the minorities' cultural features are generally restricted to the particular sphere of the cultural centers of each republic and do not move into public life, or are banned. Political parties suspected of defending ethnic rights or values are persecuted, and the constitutions of Central Asian countries do not allow for the formation of those kinds of parties because they incite ethnic hatred.

This factor, in addition to the artificial positioning of the borders, has caused, or might cause, important conflicts of every kind, and could harm human security and the human right to water. In the Amu Darya basin, the most explosive ethnic rivalries related to water resources are between Karakalpaks and Uzbeks, and between Uzbeks and Tajiks.

The Karakalpaks in Uzbekistan

The Karakalpaks are a Turkic-Mongolian people who speak Karakalpak, a language belonging to the Kipchak Turk family. These people are closely related, both linguistically and culturally, to the Kazakhs. They are Sunni Muslims.

The Karakalpaks have traditionally lived on the southern shore of the Aral Sea, in the Amu Darya delta, alternating fishing, agriculture and seasonal livestock mobility. At the beginning of the nineteenth century, the Karakalpak territory fell under the sovereignty of the Khiva khanate. In 1873, the Karakalpak were incorporated into the Russian Empire. During the Soviet era, Karakalpakstan became an autonomous region (1925), in 1932 it was an autonomous republic and in 1936 was included within Uzbekistan (37 % of current Uzbek territory).

In December 1990, the Supreme Council of the autonomous Soviet Socialist Republic of Karakalpak declared itself a sovereign state, to be effective after the citizens' approval in a referendum. After Uzbek independence, the Karakalpaks continued to be formally autonomous, and their Constitution (1993) retained a clause that enabled the possibility of deciding their independence in a referendum, but this clause has to date never been applied because of the close links of the current autonomous republic's rulers with Uzbek's President Shavkat Mirziyayev's government, by which they are actually designated.

The Karakalpaks are now a minority group in their own republic, just one-third of the population of a republic shared mainly with Uzbeks and Kazakhs. The ecologic disaster of the Aral Sea has deeply affected the region's economy and the health of the people around that sea. Karakalpakstan, one of the poorest areas in Uzbekistan, has the lowest levels of environmental and health quality; in 1998, 70 % of its rural population was considered to be poor, compared to only 10 % in the Tashkent province. This is the Central Asian region with the highest infant mortality rate: 63 infants per 1,000 live births die during their first few hours of life.

During the last decades, approximately 270,000 Karakalpaks have left the province to emigrate to Russia and Kazakhstan because of the poor economic and environmental conditions. However, there are important gas and oil reserves in the region (1.7 tcm^3 (trillion cubic meters) and 1.7 tm (metric tons) of crude).

There are some clandestine separatist minority groups, with no relevant social support, such as the Free Karakalpakstan National Revival Party, or Halk Mapi (in the people's interest), which constantly denounce the Uzbek government's lack of initiatives to solve the economic and environmental crisis suffered by the province, as well as the fact that the people do not benefit from the exploitation of its wealth in energy raw materials. They suggest independence by means of a national referendum as the solution.

UZBEK—TAJIK RIVALRY

Relations between Uzbekistan and Tajikistan are based mainly on the legendary rivalry between the Uzbeks and the Tajiks, aggravated throughout the twentieth century by Soviet nationality policies and a system of border delimitation that included several provinces with more than 1 million Tajiks in Uzbek territories. This fact, in addition to the Tajik Civil War, has caused many tense moments between them. During that conflict, the Uzbek government had an active role supplying the Uzbek groups with arms (the Uzbeks are approximately 11 % of the population) and collaborating with the troops loyal to the Tajik government against the Tajik Armed Opposition (TAO). The alliance ended when the Uzbek militia, supported by Tashkent, considered that the negotiations between Tajik president Emomali Rahmonov and the TAO were against their interests. In April 1997, a murder attempt on Rahmonov was frustrated. Rahmonov himself stated that the Uzbek government was involved. The Uzbek-Tajik general Umed Khudoiberdiev was responsible for various unsuccessful armed uprisings in Western Tajikistan, with the Uzbek government again being blamed by Tajikistan for having supported them. The episodes of war between the two countries opened a trust gap that has been difficult to bridge, with the additional problem of the human rights of Tajik and Uzbek minorities in Uzbekistan and in Tajikistan, respectively, not being met.

The Uzbeks in Tajikistan

The Uzbeks living in Tajikistan make up approximately 25 % of the population. They are settled mainly in the north of the country, near Dushanbe, and in the province of Kurgan-Tyube. Since Tajik independence, some disturbing events have taken place: an Uzbek major of the Khatlon district disappeared in strange circumstances in 1999, and during the Tajik civil war there have been multiple murders of Uzbeks in the Panj district, leading many Uzbeks to move to other districts inhabited mostly by people of their own culture.

Despite comprising 25 % of the whole country's population, Uzbeks have almost no representation in Tajik government institutions, and education in the Uzbek language is completely neglected by the Tajik government. Several western areas of the country traditionally inhabited by Uzbeks have been repopulated, following government dispositions, by Tajik families, so that certain tensions have developed between the two communities.

Relations between the Tajik and Uzbek governments deteriorated after the Tajik civil war; in 1996, two warlords from the north of Tajikistan (Ibodullo Boimatov and Mahmud Hudoberdiev) belonging to the Uzbek ethnic group, and initially allies of Rahmonov's government, tried to overthrow the latter. President Rahmonov blamed this attempt on Uzbekistan and its plans to control the country. In general, the Tajiks see the Uzbeks living in their country as a fifth column conspiring to put the country under Uzbek power.

The Tajiks in Uzbekistan

The Tajiks are the second most important ethnic group in Uzbekistan, with approximately 1.5 million of them currently living close to the Uzbek—Tajik border and in the cities of Bujara and Samarkand. The main discord between these two groups comes from the national delimitation carried out by the Soviet Union during the period 1926–1936, when the cities of Samarkand and Bujara, historically two large Tajik cultural centers, were not included in the Soviet Socialist Republic of Tajikistan, but became part of Uzbekistan. The Uzbek government tends to see the Tajik—Uzbek peoples as suspicious separatists who want to be part of neighboring Tajikistan in order to create "Greater Tajikistan" and to be connected to Islamic radical groups such as the Uzbek Islamic Movement or Hizb ut-Tahrir. According to these perceptions, since 1992, the Uzbek government has forcibly relocated a great number of Tajiks who were living in the regions bordering Tajikistan, and has even mined that border.

The Tajik language is also seen as an obstacle for Uzbek National State construction: it is not officially recognized by the Uzbek authorities, publications in Tajik are strongly censored and many books written in this language have been destroyed for being considered as a threat to Uzbek national values. Most of the Tajik cultural centers operate illegally, because they are not given the pertinent authorizations from the Uzbek authorities. The only association in favor of the Tajik culture and rights with a certain degree of activity in the early 1990s was the Samarkand Movement, led by Uktam Belmukhammedov (Banks et al. 1998); later, it was labeled illegal and its leaders were prosecuted.

In brief, the government of Uzbekistan considers the Tajiks, in general, as a potential threat to the national integrity and security of the country.

As a general rule, we could conclude that, despite the agreements signed by Central Asian countries regarding human rights, the relations between the different "non-title-holder" ethnic groups and the state are developed in a context of precariousness, lack of confidence, uncertainty and, on some occasions, fear. The psychological orientations promoted by the states towards their citizens are based on nationalist policies, lacking a real desire to legitimate the multiethnic cultural reality of Central Asia and to promote their participation in state institutions from the sphere of executive administration. Marginalization and prosecution of the "other" have been in many cases attitudes encouraged by the lack of sensibility of Central Asian states regarding the cultural diversity of their societies, and the distribution and accessibility to water resources for ethnic groups marginalized from national politics and public life, the Karakalpak in Uzbekistan being the most notorious example. We shall now examine the role of multilateral security organizations regarding human rights, the HRWS and water conflicts in the region.

THE COLLECTIVE SECURITY TREATY ORGANIZATION AND HUMAN RIGHTS

The current Collective Security Treaty Organization (CSTO), founded in October 2002, has its origins in the Commonwealth of Independent States (CIS). The CSTO is an intergovernmental military alliance signed on 15 May 1992 by six post-Soviet states belonging to the Commonwealth of Independent States—Russia, Armenia, Kazakhstan, Kyrgyzstan, Tajikistan and Uzbekistan. Three other post-Soviet states— Azerbaijan, Belarus and Georgia—signed the following year and the treaty took effect in 1994. The aim of the CIS was keeping the old links between the ormer Soviet republics, but formulated in a different way. According to the statutes of the CIS, the relations between its new members would no longer be based on the Soviet peoples' brotherhood, the government of the Soviets and a planned economy, but on democratic values, the rule of law, a free market, and interstate relations founded on the new Soviet republics' status as sovereign and independent states, as well as on their inclusion in the international community as full-member countries. The CIS's strategy to reach this aim was multidimensional cooperation with three different branches: political, economic and military, in order to integrate the former Soviet republics through the creation of a confederate organization regulating the political, economic and military relations of the countries formerly within the USSR area freely and by consensus among the countries.

The Tashkent Treaty (TT) was signed within the military branch of the CIS in 1966 and would later allow the reformulation of security relations between the new independent countries of Central Asia, the Russian Federation and other ex-Soviet republics.

CENTRAL ASIAN COUNTRIES AND CSTO: ACTORS AND CONTEXT

All Central Asian countries, apart from Turkmenistan and Uzbekistan, come under the TT. Even Uzbekistan belonged to it earlier: in June 2012, Uzbekistan sent a note to the secretariat of the CSTO, informing them that it was suspending its membership of the body. This is the second time in the history of Uzbekistan's membership of the CSTO that it has done so (the previous period of suspension ran from 1999 to 2006). Yet, even after 2006, Uzbekistan did not participate in the CSTO's initiatives in full; it did not participate in military exercises or support the formation of rapid reaction forces, and in many cases it openly contested the organization's Russian-dominated policy (for example, in 2009 it opposed the opening of a Russian military base under the aegis of the CSTO in southern Kyrgyzstan, near the Uzbek border). Initially, the former Soviet republics' becoming part of the collective security treaty was determined by several factors:

1. Ensure their sovereignty, borders and territorial integrity from Russia
   The Tashkent Treaty involved the Russian Federation's formal acknowledgment of the ex-Soviet republics as new subjects of international law, with the commitment of respecting their territory, borders and sovereignty. This was a particularly delicate question for Kazakhstan, the only country in the region that has a border with Russia, an area inhabited by a population mainly from Slavic origins.

2. The military weakness of Central Asia countries
   None of the countries had its own army. Military infrastructures, personnel and defense protocols were integrated within the Red Army, whose aim was to defend the USSR and the territory of its allies. Therefore, when the ex-Soviet republics became independent they had no efficient armies or operational capacities to cope with internal or external threats; nor did they have the financial or human capacity to create them. In this context, and with independence so recently obtained, being on good terms with the metropolis was definitely advisablea. Staying with the CSTO did not involve great political or economic commitment for the former Soviet republics; quite the contrary, in fact: it might eventually mean possible economic or military assistance from Russia.

3. Instability on its southern frontier
   The occasional terrorist attacks in Uzbekistan or Kyrgyzstan, and the rise of the Taliban in Afghanistan, put new energy into relations between the members of the Tashkent agreement, with the aim of responding to the terrorist threat in Central Asia. This was the goal of the document signed in the Armenian city of Erevan in 2001: the commitment to a joint counterterrorist fight and the creation of a counterterrorist center and a rapid reaction force, as well as to reinforce security in border areas. To reach this goal, they decided to change the Tashkent security treaty into a regional cooperation organization: the CSTO.

4. The possibility of disuading other great powers of the region such as China, which are potencials threats
Kazakhstan, Kyrgyzstan and Tajikistan had extensive borders not delimited completely by any agreement with their Chinese neighbor. The TT might be a dissuasive element for China should this government not want to negotiate.

The TT did not involve any commitment regarding security relations with the closest neighbours, and did not establish any kind of engagement with regard to security matters between the Central Asian countries themselves, all member states apart from Turkmenistan and Uzbekistan. This meant, in theory at least, that the relations between them would not be subject to any control by Russia or the organization itself.

These are some of the factors that have allowed for a certain degree of success in the securitization process led by Russia in Central Asia through the Central Security Treaty Organization (CSTO).

THE CSTO SPEECH ACT; THE TASHKENT TREATY (TT) AND RUSSIAN LEADERSHIP

The essential documents that constitute the foundations for the CSTO security concepts are the Tashkent Treaty (1992), the strategic concept and the CSTO statutes, although there are some others, such as the declaration of the member states, and the basic directives of cooperation in the military sphere.

THE TASHKENT TREATY

It is an eminently political-military treaty, according to its articles 1, 2 and 4. Its original aim was to become a new regulating principle of the political and military relations between the Russian Federation and the rest of the former Soviet republics after the dissolution of the USSR. This new scheme of relations should be based on a pact or alliance between a group of independent and sovereign states, which decide that an attack against one of them can be considered an attack on any of the others. The TT was originally inspired by the collective security principle, though it does not develop it. As some specialists show, the TT rather tries to organize and coordinate military structures and infrastructures, as article 7 of the CSTO statutes points out:

> Article 7: In order to attain the purposes of the Organization, the member States shall take joint measures to organize within its framework an effective collective security system, to establish coalition groupings of forces and the corresponding administrative bodies and create a military infrastructure, to train military staff and specialists for the armed forces and to furnish the latter with the necessary arms and military technology.

The purpose of this organization therefore suggests, as a final cooperative aim, the integration of some army bodies in joint units under a unique command, as its articles 9 and 10 state:

> Article 9: The member States shall agree upon and coordinate their foreign policy positions regarding international and regional security problems, using, inter alia, the consultation mechanisms and procedures of the Organization.

> Article 10: The member States shall take measures to develop a treaty-law base that will govern the functioning of the collective security system and to harmonize national legislation relating to questions of defence, military construction and security.

The CSTO thus plans an association of states with the aim of integrating military and political structures at the highest level. Considering the logistic, technologic, human and financial supremacy of the Russian army over the rest of the CSTO members, it is obvious that in a process of integration of military structures, Russia would play a central role in relation to the other members of the organization. In short, this organization is an attempt at reformulating the model of the old Soviet Union army, with a central command (Collective Security Council) that coordinates, organizes and decides all the actions to be carried out. In addition, it wants to recreate the old defense perimeter of the USSR.

The defensive perimeter or the border area that CSTO would be prepared to defend is thus virtually the equivalent to the old USSR's. We could then consider that the Russian Federation, whose current leadership is completely indisputable within the CSTO, wants to create a military security sphere beyond its own borders similar to its predecessor, the Soviet Federative Socialist Republic, in the context of the USSR. It moves its defense perimeter towards the borders of neighboring countries and takes responsibility for defense against possible aggression.

In the case of Central Asia, the potential aggressors for the CSTO border perimeter would be Azerbaijan (Caspian frontier), Uzbekistan, Turkmenistan, Afghanistan and China.

Oddly enough, the TT is essentially oriented towards the defense of a border perimeter against third countries, but it does not regulate security relations *between* its members; it merely suggests that possible conflicts among them should be solved pacifically, but without offering or arranging any kind of preventive measures. The only step related to a possible conflict between the TT signatories is established in its article 1: "They shall undertake to settle all the differences among them and with other states by peaceful means." An alliance against third countries does not necessarily involve cooperation to reduce the level of uncertainty between the allies themselves. Therefore, the TT is a possible response to a possible external aggressor, but it underestimates the security gaps between the members of the treaty and their respective national interests.

What Are the Main Threats Perceived in the CSTO?

The TT establishes, in its section Declaration of the States Parties to the Collective Security Treaty, signed on 10 February 1995 in Almaty, that the main political-military aim if its association is dissuasion, that it has no enemies and any state or organization can cooperate with, or be part of, the TT if they respect its goals and principles. However, while the TT does not acknowledge enemies, according to articles 1 and 8 it establishes certain limits to the security relations established by the members with other international actors, limits such as:

1. Not to join or take part in any military alliances directed against any other state party;
2. Not to conclude international agreements incompatible with the treaties adopted within the CSTO; and
3. Not to allow the stationing of armed forces and military facilities of third states in their territories without an agreement with the other member states.

The TT thus vaguely tells us what the possible threats against the security of its member states are; however, the strategic concept of the organization is much more specific. According that strategic concept, we could state that the object of the security is preferentially the state and its different attributes, such as borders, sovereignty and territorial integrity. Threats are of a transnational nature and can come from terrorist organizations, with or without a nuclear capacity, acting independently or with the support of a state; or from some other states that constitute a military threat by means of an arms race (proliferation of weapons of mass destruction, or conventional ones), subversive actions or border violations. But environmental issues or the defense of human rights are not mentioned anywhere.

CSTO AND THE DISPOSITIF

The security policies of the CSTO since its creation have been deeply linked to the implementation of some military cooperation policies:

1. The establishment of a collective army (rapid action force) through the creation of joint military structures, expecting that a current force of 4,000 personnel will ultimately have 11,000 troops distributed across three different commands: Eastern Europe, Caucasus and Central Asia.
2. To carry out joint military maneuvers with the aim of evaluating the operational and logistic capacity of the organization's members.
3. Coordination and implementation of preventive activities related to drug-trafficking, such as Operation "Channel", against the illicit traffic of drugs.
4. The creation of a joint information system between the CSTO members in cooperation with the United Nations Counter-Terrorism Strategy Programme and the CIS Counter-Terrorism Activities Centre.
5. Initiatives of cooperation with other organizations related to security, such as the United Nations Counter-Terrorism Committee, the OSCE, Eurasec, the SCO and the IMO (International Maritime Organization).
6. Control of fulfillment by some CSTO members, including the four Central Asian countries, of their international commitments regarding the dismantling and control of industrial complexes producing

components that could be used in the construction and design of weapons of mass destruction (WMD)(nuclear, chemical and biological) and the implementation of an area free of (WMD).

7. Increase military technical cooperation to harmonize the weapons, communication and logistic systems between the member countries so that the level of coordination of the troops can be improved.

The CSTO military dimension as the main guarantor of security in Eurasia is thus a priority, because the threats against the organization also have a military nature. The removal of threats is based mainly on the power of dissuasion that can be exerted by the rapid action force collective, by the logistic coordination and military operating capacity between its member countries, and by information exchange. Its primary aims are territory, border and state, while human rights or the human right to water are not considered as objects to be guaranteed and are excluded from the discourse or debates of the organization.

CSTO, Human Rights and Water Resources

The CSTO Charter contains no provisions relating to peoples' democratic will or the protection of human rights. This is a very revealing silence. None of the CSTO members stands out as a human rights champion: rather, they show a lack of commitment on this issue in practical terms. Since the early 2000s Central Asian governments have feared a possible contagion on their territory by the "color revolutions" that rocked Georgia, Ukraine and Kyrgyzstan.

Human rights might even be considered by the CSTO members as a nuisance with regard to operating capacity. The Collective Rapid Reaction Force (CRRF), was created with the purpose of repulsing military aggression, conducts anti-terrorist operations. The CSTO members not only allow the CRRF to act in a wide range of (both external and internal) crises but also removed the need for full consensus prior to initiating its operations. In this regard, it is noteworthy that, in fact, the CSTO rapid reaction (or peace-keeping) forces would most likely be used to ensure the stability of the political regimes in the territories of the CSTO member states.[13]

The security concept of this organization, based exclusively on a military perspective and at the service of sovereignty, borders and specific interests of the states, is reducing opportunities for transboundary cooperation in the Amu Darya. This organization has no programs or activities related to Central Asian water resources and environment, but the region lies completely within its strategic ideals because of the borders shared by its members and their nearness to the Amu Darya river. This river represents, according to the CSTO, a traditional defensive frontier, with potential aggressors such as Turkmenistan, Afghanistan or Uzbekistan, and potential allies, such as Kazakhstan, Tajikistan and Kyrgyzstan. The river is thus seen as a possible confrontation scene rather than a cooperative border. The Amu Darya is considered by the CSTO as an area for its own security, and, consequently, any border, territorial or sovereignty conflict, as well as any agreement that means a cessation of sovereignty or territorial changes related to the river's water resources management, could be interpreted as a threat to the security of the CSTO.

The Shanghai Cooperation Organisation (SCO) and Human Rights

In the 1990s, China, Russia and the Central Asian countries held several summits, whose main targets were dealing with new border delimitations, promoting transboundary trust, and cooperation to deal with the three most relevant threats to the region from their point of view: extremism, separatism and terrorism.

According to the articles 1 and 2 of the Shanghai Cooperation Organisation's (SCO's) Constitution, these are the principles that must govern the relations between its members:

1. The members of the SCO must strictly observe the principles and purposes of the United Nations Charter: mutual respect for independence, sovereignty and territorial integrity, not intervening in internal affairs, and refraining from the use of threat or force among its members, as well as the peaceful resolution of their problems and not unilaterally or by military means.
2. To strengthen friendship and trust between the member states, promoting efficient cooperation in the fields of politics, economy, commerce, science and technology, culture, education, energy, communications and environment, among others.
3. To promote a new international order based on democratic transparency and political and economic rationality. To cooperate to establish a new fair and rational international order.

Some of the above-mentioned ideas are inspired by collective security values and they refer to the respect and acknowledgement every member of any organization or treaty must observe toward the sovereignty and territorial integrity of its neighbors, and its will to solve possible conflicts peacefully. However, the SCO does not automatically guarantee any help or defensive action in the case of an attack by a third country, a crucial idea in the concept of collective security, as it is established by the CSTO or NATO. There are no perspectives of joint military defense in case of a hypothetical attack against an SCO member, though article 6 of the Treaty on Long-Term Good-Neighborliness, Friendship and Cooperation between the Member States of the Shanghai Cooperation Organisation, establishes that some measures might be taken after consultation among its members, but with no further specifications.

The better-developed aspect within the multilateral cooperative works of the SCO, aside from the demilitarization measures included in the "Treaty to Implement Mutual Trust in Border Areas" (1996), has been the cooperation against extremism, separatism and terrorism.

THE COOPERATION AGAINST THE THREE EVILS AND THE HUMAN RIGHTS

In June 2001, the SCO members signed the socalled "Shanghai Convention on Combating Terrorism, Separatism and Extremism." The goals of this convention were to establish a basic document, with the general agreement of its members, intended to combat terrorism, separatism and extremism in the region. This document established the foundations of cooperation among the organization's states.

With the aim of cooperating permanently in this field, the member countries decided to create a regional center for anti-terrorism in Tashkent (June 2004) (RATS—Regional Antiterrorism Structure). These are some of the functions of the center:

(a) To establish a common database for the SCO members.
(b) To contact other bodies or states devoted to combating terrorism.
(c) To prepare reports and dossiers related to activities on terrorism, separatism and extremism.
(d) To cooperate in the detection of terrorist attacks within the SCO territory.

(e) The harmonization of anti-terrorist legislation among its members.
(f) To share lists of the most wanted terrorists.
(g) To train intelligence services personnel.

The activities of the center are focused basically on assisting the fight against terrorism in the region, because it lacks operational or military capacity.

It is difficult to evaluate its efficiency and level of success, but according to its own sources, the center has helped to arrest at least 15 terrorists. Its main target is to dismantle or to eliminate some opposition groups such as the Uygurs, the Chechens and organizations such as the Islamic Uzbek Movement, Hizb ut-Tahrir.

In addition to their cooperation in antiterrorist matters, the SCO countries periodically carry out a series of joint military maneuvers with the purpose, according to the organization itself, of combating and eliminating terrorist groups. While some of these maneuvers, especially the so-called Mission of Peace (2005), have been considered by some experts as a warning or a threat of a possible foreign power intervention in the region because of the participation of bomber squadrons and heavy artillery.

On some occasions, NGOs such as Human Right Watch have denounced SCO members for using the organization as cover to eliminate some dissident and opposing political groups instead of terrorists; considering, for example, the Andijan massacre in May 2005 as an antiterrorist operation.

The international, non-governmental advocacy group Human Rights in China (HRIC) released in March 2011 its report on Counter-Terrorism and Human Rights: The Impact of the Shanghai Cooperation Organisation. The white paper's authors expressed concern about the SCO's counterterrorism policies and practices.

THE OSCE AND HUMAN RIGHTS

The basic document that offers us the values and principles of the OSCE is the Helsinki Final Act, which includes the following ten items:

A. Sovereign equality, respect for the rights inherent in sovereignty . . .
B. Refraining from the threat or use of force . . .
C. Inviolability of frontiers . . .
D. Territorial integrity of states . . .
E. Peaceful settlement of disputes . . .

F. Non-intervention in internal affairs...
G. Respect for human rights and fundamental freedoms, including the freedom of thought, conscience, religion or belief...
H. Equal rights and self-determination of peoples.
I. Co-operation among states.
J. Fulfillment in good faith of obligations under international law.

These are the ten rules that must govern the relations between the states and the values to be defended—human rights and free determination of the people, among others. However, the activities of this organization in Central Asia have been restricted to offering help with electoral processes as well as courses, assessment and actions related to:

1. The implementation of human rights and democratization.
2. Problems derived from migratory flows (trafficking of human beings).
3. Prison policy and judicial procedures.
4. Gender equality.
5. The promotion of non-governmental mass media and organizations.

Meanwhile, the role of the OSCE has been reduced to a series of technical assistance or assessment actions related to training and public awareness on human rights, electoral processes, transparent economic management, and environmental problems; the organization has not established cooperation forums or multilateral action regarding water resources and development of the HRWS.

In spite of its multidimensional perspective the organization maintains towards security, similar that of the Shanghai Cooperation Organisation, it has not yet become a discussion or cooperation forum for water-related questions, neither in Central Asia nor among the Amu Darya's neighboring states; and it considers the transboundary watercourse of the Amu Darya to be too troubled an area even to cooperate in matters of traditional security. This is a consequence of the disagreement within the organization regarding how security matters must be dealt with in Afghanistan, one of the Amu Darya's neighboring countries. We shall now examine this situation.

In April 2003, the OSCE Permanent Council admitted Afghanistan as a new partner of the organization in the "interest of Afghanistan in establishing relations with the OSCE, based on the exchange of points of view and information about several questions of mutual interest, because the country shares borders with three states of the organization (Turkmenistan, Uzbekistan and Tajikistan)" as well as common interests in matters of security with its neighbors and other OSCE members.

After its incorporation as a partner, Afghanistan began to participate in OSCE activities and meetings, though not as full member. One of the activities was the presence of OSCE observers at the Afghan elections in 2004 and 2005.

In November 2007, at the Madrid Summit, the OSCE acknowledged the need to strengthen its cooperation with Afghanistan and other security organizations, such as CSTO and NATO, with the purpose of avoiding a duplication of its efforts in several priority areas: border management and security, police training, and the fight against drug trafficking.

In October 2008, the OSCE published a program of additional activities (Programme of Activities on the Implementation of the MC Decision 04/07) with the aim of carrying out the plans in cooperation with Afghanistan. This document did not receive the necessary agreement, however, and at the time of writing is still awaiting approval.

The next summit took place in Astana in December 2010. This meeting became a milestone in the history of the organization, because it was the first time that a former Soviet country had organized a summit, and it was held in Central Asia, a geographic framework far away from usual important events. The Kazak presidency of the OSCE stirred up many expectations about the OSCE's commitment with Afghanistan, because all the neighboring countries see the current situation in this country as the main threat to their security.

However, the summit did not mean any significant advance in terms of cooperation of the organization with Afghanistan, though in its final statement it manifested once more its concern regarding the situation in this country, and the need to make efforts to gain an adequate level of cooperation regarding the statement:

> We admit that security in the OSCE area is indissolubly linked to the adjacent areas, especially in the Mediterranean and Asia. We must therefore improve the level of interaction with our partners in terms of cooperation. In particular, we underline the need to contribute efficiently, according to the capability and the national interest of every state involved, to the collective international labor in order to promote a stable, independent, prosperous and democratic Afghanistan. (Organization for Security and Co-operation in Europe 2010)

In short, and as the organization members and their documents acknowledge, the OSCE has not to date been able to reach any agreements on political-military cooperation programmes between its members with respect to Afghanistan. And, according to the information provided by the OSCE website, it has not carried out any initiatives regarding cooperation programs regarding resources between Afghanistan and its Central Asian neighbours, showing its incapability to implement regional cooperation in general terms because of the rivalry or lack of cooperation of its most prominent members.

NATO and Human Rights

According to the document "Backgrounder Partners in Central Asia", the partnership between NATO and Central Asian countries must be ruled by:

> Partnership values: Partnership is about more than practical cooperation—it is also about values. When partner countries join the Partnership for Peace, they sign the PfP Framework Document. In doing so, partners commit to respect international law, the UN Charter, the Universal Declaration of Human Rights, the Helsinki Final Act, and international disarmament and arms control agreements; to refrain from the threat or use of force against other states; to respect existing borders; and to settle disputes peacefully. (Bacquelaine 2003)

However, the Central Asian countries' commitment to human rights is very weak, and NATO's determination to force these countries to respect them is no stronger, as the vice-president of the NATO Parliamentary Assembly, Daniel Bacquelaine, explained in a report of the NATO Parliamentary Assembly Political Committee:

> In addition, the human rights records of Central Asian regimes have made it difficult for the governments of NATO member states to engage in closer co-operation. At its core, NATO is a political-military Alliance for the collective defense of its member countries which also includes projecting stability outside its territory. NATO's ability to influence the governments of the region to improve their records on democratization, tackle corruption and criminal networks is limited

This is also the main idea in the complaints reported by NGOs and international bodies. According to reports and statements from the representative of the NGO Human Rights Watch for Europe-Asia: "'Security concerns linked to Afghanistan are important, but continuing rights abuses in Central Asian states also present a threat of long-term instability in their own right,' said Hugh Williamson, Europe and Central Asia director at Human Rights Watch."

In accordance to the World Report elaborated by Human Rights Watch in 2014, the latest events related to human rights violations are these:

- In Kyrgyzstan, ill-treatment and torture remained pervasive in places of detention in 2013. Shortcomings in law enforcement and the judiciary contribute to the persistence of grave abuses in connection with the ethnic violence in southern Kyrgyzstan in June 2010, and to the harassment and abuse of lesbian, gay, bisexual and transgender (LGBT) people. Human rights defender Azimjon Askarov remains wrongfully imprisoned.
- In Kazakhstan, authorities continued to crack down on free speech and dissent through the misuse of overly broad laws such as "inciting social discord" and through the closure and suspension of independent and opposition newspapers. Government critics such as opposition leader Vladimir Kozlov remain imprisoned following unfair trials. There were several cases of forced psychiatric observation or detention, in violation of rights.
- In Tajikistan, President Emomali Rahmonov, in power since 1992, was re-elected for a fourth term in office in November 2013. The only independent candidate, Oinihol Bobonazarova, was forced to exit the race prematurely after her supporters were intimidated. In the lead-up to the election, authorities widened a crackdown on freedom of expression, imprisoned opposition leaders, shut down a leading NGO, and stepped up efforts to extradite political opponents. The authorities exercise tight controls over religious freedoms. Torture remains common in places of detention.
- Authorities in Turkmenistan released several political prisoners and adopted several laws that some international partners have hailed as "reform." However, Turkmenistan remains one of the world's most repressive countries. It is virtually closed to independent scrutiny. Media and religious freedoms are subject to draconian restrictions, human rights defenders and other activists face the constant threat of government reprisal, and the government continues to use imprisonment as a tool for political retaliation.

- Uzbekistan's human rights record remains abysmal. Dozens of civil society activists remain behind bars for no reason other than their human rights work, along with thousands of peaceful religious believers. Many of these individuals have been subjected to torture, which the United Nations Committee Against Torture (CAT) found in November 2014 to be "systematic." The Uzbek government forcibly mobilized nearly 2 million adults and children during the autumn cotton harvest to pick cotton in abusive conditions for little to no pay.

The first conclusion, or consequence, we can establish is that multilateral security organizations' policies in Central Asia and the Amu Darya basin, in terms of human rights, are very limited or non-existent, despite the fact that NATO, the OSCE and the SCO proclaim human rights as a value to be defended; though, as in the case of the SCO, only when it does not contradict the internal legislation of its members.

One of the reasons explaining the lack of commitment of these multilateral security organizations to implement human rights and the human right to water in the Amu Darya basin is, among others, that some of them, such as the CSTO, NATO and the SCO, have given priority to policies of military and police cooperation to solve what they consider the main threat in the region: terrorism and Afghan insurgence.

THE MACROSECURITIZATION OF SECURITY IN CENTRAL ASIA

According to Ole Waever, to securitize an issue is "to demand the mobilization of maximum effort" (1995, p. 53), and justifies the "use of extraordinary measures to handle them" (Buzan et al. 1998, p. 21). The September 11 attacks in Washington and New York in 2001, and the subsequent invasion of Afghanistan by NATO troops and the International Security Assistance Force (ISAF), meant an opportunity to begin a process of readjustment of the international order led by the US Bush Administration, using the rhetoric of the values of war against terrorism, making it the "dominant macrosecuritization issue around which US, Western and even Russian, Chinese and Indian foreign policy [can] be coordinated" (Wendt 1999. p. 272). In effect, that terrorist threat rhetoric was reformulated by China and Russia to justify new police measures in their own territories and their foreign relationships. The Central Asian countries also adopted this rhetoric, even though it has almost nothing to do with terrorism, as shown in the Andijan massacre (2005) in Uzbekistan, or the bunkerization and expansionism policies of the

Uzbek borders against its Turkmen, Kyrgyz or Tajik neighbors, apparently linked more to border control and access to grazing or cultivation lands, than with the infiltration of terrorist groups.

The NGOs also suffered the consequences of these actions. According to reports from the International Helsinki Federation for Human Rights (IHF), the Central Asian countries established new legislative measures restricting theNGOs' defence of human rights and liberties, especially those funded by foreign institutions. Such a restriction of liberties was carried out for the sake of national security and the need to prevent extremism.

The vagueness of macro-securitization allows the threat of terrorism to be malleable, adaptable, and thus broadly applicable on an international level. Macro-securitisations, such as the Cold War and terrorism, allow international relations to impose order and hierarchy to reduce security concerns; or bundle other security threats into the overarching macrosecurity concern (Wendt 1999, p. 257).

Another factor that strengthens the trend towards the macrosecuritization of the region—and the monopolization of the decision-making process with regard to the concept of security and the policies to be applied—is the lack of democratic governance of international security organizations. According to P. Mayer (2008): "In general terms, the international organizations have tried to keep the gates closed to the public. In the field of international security, the stakes for states are extraordinarily high. However, as much of international security policy has shifted from alliance formation towards conflict prevention and peace building, the collaboration of the civil society has become desirable, if not inevitable". P. Mayer describes some of the keys to implementing participation mechanisms between international institutions/ organizations and civil society, with the aim of legitimating them for their audiences:

> If organized civil society has the opportunity to participate in international governance, it may act as a transmission belt between international organizations and an emerging transnational public sphere. This transmission belt might operate in two directions: first, civil society organizations can give voice to citizens' concerns and channel them into the deliberative process of international organizations. Second, they can make internal decision-making processes of international organizations more transparent to the wider public and formulate technical issues in accessible terms.

In this sense, transnational advocacy networks can play a key role in promoting participation and transparency in international organizations operating in the field of security.

THE TRANSNATIONAL ADVOCACY NETWORKS (TANS)

A term coined by Margaret Keck and Kathryn Sikkink, transnational advocacy networks are coalitions of grassroots movements and organizations that exert international influence through soft power strategies. The TANs should participate in the decision-making processes of international security organizations, and in the process of constructing the community of security around water. In this context, civil society organizations have had a fundamental and growing role in the protection of common public goods such as environmental and human rights to water and the promotion of sanitation. The role of advocacy networks is to connect civil society actors with states and international organizations, thus multiplying channels of access for such actors to the international system. One of its major exercises is to conduct exchanges of strategic information in order to place the issues into categories that may persuade, press and influence actors in decision-making (Keck and Sikkink 1998). By creating new issues and placing them on both international and national agendas, providing crucial information to actors, and most importantly by creating and publicizing new norms and discourses, transnational advocacy groups help in restructuring world politics (Schmitz and Sikkink 2002, p. 306).

What distinguishes advocacy networks from other existing transnational associative categories—transnational social movements, coalitions, transnational non-governmental organizations (TNGOs)—is the plurality of actors composing it, and the role played by domestic and international NGOs. These, as central actors, would be the promoters of voluntary associations through formal or informal relationships between members of the network. According to Keck and Sikkink (1998), in addition to the NGOs, the main actors that make up the advocacy network are: local social movements, foundations, media, churches, trade unions, international and regional organizations, and parts of the executive and the parliamentary body of the states when they join the cause promoted by the network.

More actors need to become involved in NATO decision-making and share its values to perceive it as legitimate, because the outcomes of the organization concerns audiences of various kinds, often far away. Civilian actors such as NGOs have been highlighted here as essential players for

NATO in its attempts at achieving a comprehensive approach. They are also an audience that assesses NATO's actions. These new audiences have previously not been considered to be important theoretically in the security governance literature.

In the specific case of Central Asia it would also be advisable to include "certain forms of traditional network or kinship structures as an intrinsic part of civil society that cannot be ignored or overwritten, and that including them within the definition of civil society makes more sense for certain non-Western contexts" (Giffen and Earle with Charles Buxton 2005, p. 54). This is the case with traditional institutions such as mahalla, avlods, jamoat, uru, ru, and clan and tribal networks.

While examining the nature of civil society in Central Asia, it can be concluded that the influence of traditional and informal institutions on the social and political life of Central Asian society is very powerful. Kyrgyz uru, Kazakh ru and Tajik avlods have similar functions in implementing social activities in rural areas. These three informal institutions play significant roles in identity building, and the organization of social activities in cultural areas. While uru, ru and avlods do not carry official legal status in the Kyrgyz, Kazakh and Tajik governmental systems, they have a strong presence in the civic and social life of these countries. Uru, ru and avlods are seen as a part of cultural heritage. Analyses of Soviet policy towards informal and traditional institutions reveal a twofold approach during the Soviet time. On one hand, for the Soviet Union it was important to develop formal institutions through communist ideology, education and modernism. On the other hand, traditional institutions such as the mahalla system were also seen as complementary for institution building in rural areas of Central Asia. After the collapse of the USSR, post-Soviet Central Asian leaders also displayed very careful approaches to informal institutions. If, in the case of Uzbekistan and Tajikistan, mahalla became an integral part of their society, Kyrgyz, Turkmen and Kazakh societies incorporated clan and tribal networks for political and civic participation.

To conclude, the construction of a security community in Central Asia demands the broad participation not only of formal and informal actors, and at every level (local, national, international), but it must also be a participation with a common purpose: to guarantee the human right to water and sanitation as the cornerstone for other human rights and the process of emancipation of its citizens.

WHERE IS THE MANAGEMENT OF WATER RESOURCES HEADING IN CENTRAL ASIA? HUMAN RIGHTS AND WATER

The current national laws of the Amu Darya basin countries are not aimed towards incorporating the human right to water and sanitation. Tajikistan and Afghanistan have acknowledged this right, but they have not implemented any mechanism to develop it. The Amu Darya basin countries, at least in theory, are planning to develop integrated water resources management (IWRM) on a national scale.

THE IWRM

In Central Asia, the IWRM is promoted through the European Union Water Initiative (EUWI) and the so-called National Policy Dialogues (NPDs). In particular, NPDs on integrated water resources management are supported financially by the EU, the Governments of Denmark, Finland, Germany, Norway and Switzerland, and the O S CE.

It is difficult to evaluate at this time whether this initiative has been successful, because the IWRM has been incorporated only recently into Central Asian legislation and there is apparently no notice regarding the actual implementation of legislative packages approved by the region's governments. Another source of distrust about the possible success of this initiative on integrated water resources management in the Amu Darya basin is that the EUWI negotiates different programs with each of the riverside countries, but not encouraging a joint implementation of them. Meanwhile, some key countries in the region, such as Uzbekistan or Afghanistan, are not part of this initiative, plus there are serious doubts about the funds with which these programs are provided. The next table shows us the different legislative packages that some riverside countries have committed to putting into practice, with Kyrgzstan, Tajikistan and Turkmenistan being the states with the highest degree of engagement in cooperation with the EUWI.[14]

The IWRM is defended by some professional water sectors, the private initiative represented by the Global Water Partnership, bodies and international organizations such as the Organization for Security and Co-operation in Europe (OSCE), the World Health Organization (WHO) and the World Bank (WB) or the European Bank for Reconstruction and Development (EBRD). Some of its principles are included in the UNECE Convention on the Protection and Use of Transboundary Watercourses and International Lakes (Water Convention), the UNECE/WHO—Europe Protocol on Water and Health, the EU Water Framework Directive (WFD) and other relevant documents.

The IWRM's most common definition has been provided by the Global Water Partnership: The Integrated Water Resources Management (IWRM) is a process which promotes the coordinated development and management of water, land and related resources in order to maximize economic and social welfare in an equitable manner without compromising the sustainability of vital ecosystems. It includes four general principles:

> Principle No. 1—Fresh water is a finite and vulnerable resource, essential to sustain life, development and the environment
> Principle No. 2—Water development and management should be based on a participatory approach, involving users, planners and policy-makers at all levels
> Principle No. 3—Women play a central part in the provision, management and safeguarding of water
> Principle No. 4—Water has an economic value in all its competing uses and should be recognized as an economic good.

These four principles aim to be a water resources management guide. This system of management is based on Bernstein's (2001) definition as "liberal environmentalism", Goldman's (2005) as "Green neoliberalism", and Bakker's (2003) as "market environmentalism". The liberal discourse has a Roman law consideration of water as an economic good susceptible to being an object of property and, particularly relevant, of economic and commercial traffic. The aim of the commercialization of water is to obtain the best possible service at the most economical price, maximizing its uses: the market and the competition will achieve a higher degree of effectiveness. People are seen as consumers or customers. The best protection of the rights and interests of the consumer/customer will be the free market and the mechanisms of private law.

The IWRM has some positive aspects: civil society participation systems; the inclusion of women in decision-making processes; a certain degree of concern about the environment and the quality of water resources; and a multisector, integrated vision of water resources (energy, human consumption, agriculture). However, this model does not guarantee access to water and sanitation as a human right. The IWRM considers water to be basically a commodity rather than a common good.

This involves the relations established by the different uses of water (human, agricultural and energy) must be defined mainly by the market rather than by human rights. The regulating or administrative principle of water resources must be the relations between providers and consumers; and companies, with or without the support of the states, must guarantee the functioning of the market. The IRWM sees water as a commodity, focusing on the technical management of the product and its subsequent sale with the aim of gaining the maximum benefit at minimum cost. Meanwhile, sustainability is perceived in economic rather than environmental terms. For example, the European Union Framework Directive based on the IRWM points out in article 9 that the member states will take into account the principle of cost recovery of the costs related to water services, including environmental ones and those linked to the resources, considering the economic analysis carried out and particularly according to the idea that who contaminates must pay. In my opinion, this principle does not prevent pollution or third-parties damages, but rather it establishes that pollution is sustainable provided that those who contaminate can pay compensation for it—of course, at the expense of the environment. The implementation of the human right to water and sanitation involves water sources being managed so that they are protected in their form of potable water, and enough safe water is available for personal and domestic use (cooking, drinking, personal and domestic hygiene) as a priority over agriculture or industry. It also means that everybody has the right to be informed and to participate effectively in the elaboration of water policies. A perspective based on human rights will force all the title holders to be responsible for the compliance with human rights (and it includes the agencies in charge of the river basins).

IWRM versus HRWS

The main discrepancy between these two concepts is probably based on the IWRM's consideration of water as an economic value, and as a social and cultural good in the case of the HRWS. However, both models can be compatible. The IWRM and the HRWS share similar principles: a participative perspective, non-discrimination, responsibility, sustainability and transparency.

An approach to water resources management based on human rights guarantees that the aim will be, in first place, to accomplish all the commitments on human rights and to establish the legal requirements to ensure universal access to water and sanitation (access to water is thus no longer seen as a need but as a right). The HRWS provides a precise, exact, verifiable and legal definition of the idea of acceptable access to water and sanitation.

On the other hand, the HRWS will benefit from the IWRM by using previously tested planning tools and examples that incorporate all the aspects related to water management. An approach focused exclusively on water and sanitation, and whose only concern is the immediate need to provide access to them, is not likely to be ready to give a good response to the issue of the human right to water and sanitation, especially with regard to environmental and economic sustainability, consequently also endangering social sustainability.

NOTES

1. Progress on Drinking Water and Sanitation—2014 update, p. 3, fig. 4. WHO Library Cataloguing-in-Publication Data.
2. International Crisis Group Europe and Central Asia Report No. 233, 11 September 2014. Water Pressures in Central Asia p.13.
3. Poyry, "Environmental and Social Report (Draft) Volume I, Rogun HPP," Ref. No. 9A000304.01, 17 June 2014. https://www.worldbank.org/con tent/dam/Worldbank/Event/ECA/central-asia/ESIARogunVolIII_eng. pdf. (accessed 18 June 2014), hereinafter "Draft ESIA," ch. 2.
4. Swatuk, Larry A. (2012): *Environmental Change, Natural Resources and Social Conflict*, (Palgrave Macmillan, 2012): Water and Security in Africa: State-Centric Narratives, Human Insecurities, p. 125.
5. Guliyev, Farid (2005): Post-Soviet Azerbaijan: Transition to Sultanistic Semi authoritarianism? An attempt at Conceptualization. *Demokratizatsiya: The Journal of Post-Soviet Democratization* 13(3): 393–435.
6. Modern Nondemocratic Regimes, in Problems of Democratic Transition and Consolidation (Baltimore: Johns Hopkins University Press, 1996).
7. Kirkpatrick, Evron M. (1989): Homo Sovieticus. *World Affairs*. Volume: 152. Issue: 2. p.104.
8. Roy, O. (2000): *The New Central Asia: The Creation of Nations*. (New York University Press), London, p. 120.
9. Ibid., p. 110.
10. East European and Former USSR Resources. Genocides and Ethnic Cleansings of Central and East Europe, the Former USSR, the Caucasus and Adjacent Middle East—1890–2007. http://filebox.vt.edu/users/ras musse/slavic/genocide.html.

11. Payin, Emil (1992): *Cultural survival. Population Transfer: The Tragedy of the Meskhetian Turks.* 16.1 (Spring 1992) After the Breakup: Roots of Soviet Dis-Union. http://www.culturalsurvival.org/ourpublications/csq/article/population-transfer-the-tragedy-meskhetian-turks.
12. Roy, O. (2000): *The New Central Asia: The Creation of Nations.* (New York University Press), London, p. 275.
13. Kembayev, Zhenis (2014) *The Emerging Eurasian Union: Problems and Perspectives.* Series editor: Thomas Kruessmann. ISBN 978-3-9503853-1-1 18. July 2014. Public Policy Research Paper no. 3.
14. OECD. http://www.oecd.org/env/outreach/partnership-eu-water-initiative-euwi.htm.

CHAPTER 6

Conclusions

Abstract In this chapter, the author draws final conclusions on the security policies of international security organizations in the area related to water in each of the dimensions of human security (environmental, economic, social and political) and possible action guidelines to mitigate water conflicts in the region with reference to the Human Right to Water and Sanitation (HRSW). The creation of a security community focused on HRWS could be an answer to the current lack of cooperation on water resources in Central Asia, showing the advantages of this resource when it is shared and not hoarded (zero-sum games), though this new culture of water, based on the HRWS, would obviously demand, as Immanuel Kant might have said, a learning process, a process of construction of a new common identity built around an indispensable element of human life and the welfare of people in Central Asia: water.

Keywords Water insecurity · National construction · Extraordinary politics · Community security · Civil society · Empowerment

In accordance with the *Water GEF Fund Environment Global Review* data analyzed in previous chapters, the situation of the human right to water in Central Asia is especially serious: more than 19 million people have no access to drinking water, and more than 30 million do not have such access

© The Author(s) 2017   153
M.Á. Pérez Martín, *Security and Human Right to Water in Central
Asia*, Security, Development and Human Rights in East Asia,
DOI 10.1057/978-1-137-54005-8_6

guaranteed. According to the World Health Organization (WHO), there are more than 26 million people (no data is available about Turkmenistan) have no adequate access to drinking water in Central Asia. At the same time, the most recent data on the observance of the Millennium Development Goals (MDGs) elaborated by the WHO and UNICEF say that, in recent years, the number of people with access to drinking water and sanitation has dropped by 1 %, a percentage that in this region means more than half a million people. In addition to the HRWS figures, there are 4 million people malnourished in Central Asia, and 2.2 million in Afghanistan; meanwhile, millions of people, especially in Tajikistan, Afghanistan and Uzbekistan, suffer from energy poverty. However, Central Asia is a region rich in water and energy resources in comparison to other areas around the world.

WATER, ENERGY AND WEALTH

Central Asian countries are among the biggest consumers of water in the world in relation to the sizes of their economies and populations. All these data allow us to conclude that the region is relatively rich in water resources compared with others. It is also rich in energy resources, as we saw in previous chapters. However, as was mentioned, there are millions of people without water, with a deficient quality in many areas, with no sanitation, malnourished and with an inadequate electricity supply.

The major issue in Central Asia is not insufficient water quantity, but rather imbalanced water consumption against average annual flow generation. While Afghanistan, Tajikistan and Kyrgyzstan are the most upstream countries in the Aral Sea basin, providing about 80 % of the annual flow, water withdrawals for these three countries total 17 %. The picture for the downstream states (Uzbekistan, Kazakhstan and Turkmenistan) is exactly the opposite: Uzbekistan, which generates about 8 % of the flow, withdraws about 52 % of the total water, followed by Turkmenistan, 20 %, and Kazakhstan, 10 % (Allouche 2007).

Water mismanagement in some of the Central Asian states and the over-exploitation of water by agriculture and industry, the biggest water consuming and polluting sectors in the region, are threatening the human right to water and sanitation and, consequently, the economic right of individuals to enjoy an adequate standard of life. This situation of scarcity and danger for human security in Central Asia is expected to intensify in the following years, given the expectation of an increase in water consumption caused by strong demographic and economic growth. The highest population growth rate will be demonstrated by Uzbekistan, where the number or residents is expected to rise by almost 30 % by 2050—over 11 million people. The population of Tajikistan will increase by almost 3.8 million people. By 2050, the population of Turkmenistan and Kyrgyzstan will rise by 1.8 and 1.3 million people, respectively.[1,2]

During the period 1957–1980 the glaciers in the Aral Sea basin lost 20 % of their ice (EDB and EB IFAS 2009). The melt water from glaciers in Tajikistan contributes in an average year 10–20 % of the runoff of large rivers in the region. In hot, dry years the contributions to certain rivers can reach as much as to 70 % in the summer period (EDB and EB IFAS 2009). The glaciers of Tajikistan decreased by 20–30 % overall during the twentieth century. In Afghanistan the decrease has ben as much as 50–70 %. This glacial melting may increase the runoff to the rivers in the region in the short term, but the long-term effect of the depleted glaciers will, however, be reduced runoff (EDB and EB IFAS 2009).

The runoff of the Syr Darya river is not expected to exceed the natural variations in any of the scenarios for the period up to 2030. The calculated models for Amu Darya show a reduction of water by 5–8 % until 2030. When looking at the period until 2050, a reduction in both rivers is expected (these scenarios have excluded the precipitation variable, which adds to the uncertainty of the predictions).[3]

Therefore, the various uses and types of water consumption will be affected, thus aggravating vulnerability, and the environmental, economic, political and social risks related to water. The most important challenge for peace and human security in Central Asia will be dealing with the diversity of threats and dangers derived from water mismanagement, and the lack of cooperation between its states and organizations or international initiatives in terms of water resources.

Risks and Threats

As was mentioned in previous chapters, there exist a wide range of threats to human security in the region: environmental (desertification, intensive use of insecticides and fertilizers harmful to the environment, global warming, nuclear pollution risks—uranium waste dumps—or pollution coming from extractive or military industrial complexes); economic (corruption, unemployment, labor exploitation, migration and a strong growth in the informal economy (illegal trafficking networks); political and sociocultural (ethnic conflicts, terrorist groups and drug trafficking, autocratic governments, lack of true guarantees to exercise individual and collective human rights, including the human right to water—HRWS).

Most of these threats are of a transnational nature, and are the cause or consequence of water resource management models or are related to them, directly or indirectly:

A. Threats to environmental security are a consequence of considering water merely as an economic good or another production factor rather than seeing it as a key foundation of ecosystems.

B. Threats to economic security are caused by a management model based on the overexploitation and contamination of water resources, because of the prevalence of certain activities involving the consumption of large amounts of water (the cotton sector), the maximization of farming production based on new irrigated lands, and the use of a cheap and unskilled workforce as well as, in some cases, women and children in slave-like conditions.

C. Threats to political and social security are linked to political decisions made arbitrarily, in a non-transparent way and not agreed by consensus. They are related to the way that water is managed (how much, when, how, who, and for what in terms of consumption) and consider water as a power resource, so that some groups or actors benefit at the expense of others, leading to a precarious and troubled context among the different actors. However, water problems—when combined with poverty, social tensions, environmental degradation, ineffectual leadership and weak political institutions—contribute to social disruptions that can result in state failure.[4]

The final consequence of this mixture of factors deeply affects, and will continue to affect, the amount and quality of water resources, because:

1. The above-mentioned environmental threats affect the amount and quality of water resources, and therefore their economic exploitation and human consumption.

2. The overexploitation of water resources and their use in certain economic activities, such as the cotton production or the extractive-mining sectors, also concerns the amount and quality of water resources available for other uses.

3. The political decisions related to the way that water is managed, made in a non-transparent way and without agreement, create conflicts—not necessarily violent—among different actors, affecting the amount and quality of the available water resources (corruption) and the willingness of states to cooperate with other states over water management issues.

Meanwhile, the responses of Central Asian states to these threats, as well as the international cooperative reactions previously described, have not been very encouraging.

The Central Asian countries' compliance with international agreements on environmental issues is at a minimum. Turkmenistan and Uzbekistan show the lowest degree of commitment to this, as they are not part of the conventions related to the fight against cross-border pollution or contaminating substances used in agriculture. On the other hand, Kyrgyzstan, Kazakhstan and Tajikistan have signed those agreements, and both Kazakhs and Kyrgyz have published a list of banned products, even though their commitment on other aspects is rather weak. Regarding the Convention on Climate Change, all the states have ratified it, but only Kazakhstan has signed one of the two annexes which put the compliance with this agreement into operation. The reasons for the low degree of involvement among the Central Asian states on environmental issues are a lack of funding, specialized human resources or environmental values and awareness, as well as ineffective control and supervision methods or the costs that some economic sectors are not willing to support. With regard to the elimination and management of the uranium waste dumps spread over the entire region, the efforts of the international community have not succeeded as expected, because of the high costs of clean-up and maintenance, and the lack of donors to finance these. The nuclear-weapon-free zone might have been the multilateral response to eliminate those dumps, but the agreement is focused on the commitment of Central Asian states not to build and deploy nuclear weapons in their territories. At the same time, the cooperative attempts developed by the Fund for Saving the Aral Sea have been relatively important locally, but have failed to become a multilateral cooperation forum because of the lack of coordination among the different actors involved in cross-border water resources management and the lack of political will among the states. Regarding the possible environmental impacts of the industrial military complexes still working in Kazakhstan, I have not found any information about them to date.

In the economic field, the efforts of the Central Asian states are aimed only at the growth of their macroeconomic figures and the maximization of raw materials production, while Central Asian societies are, in general, poorer every day and have fewer employment options and fewer possibilities of social assistance, education and professional training. The general reasons are linked to the states' loss of tax capacity, non-transparent privatizations, state interventionism on prices, high inflation rates, drastic cuts in health, education and research, and the inadequate distribution of public resources (as a result of corruption). The consequence of all this has been a progressive growth in the informal economy, and migration and crime, especially connected with drugs.

Nowadays the increase of productivity in the farming sector, the largest one in the region in terms of employment, is not based on innovations in technology, or business management, but rather on the non-remunerated, or barely remunerated, labor of women and children. If the process of the deterioration of the amount and quality of water resources continues, either caused by human actions (consumption increases) or environmental reasons (droughts, global warming, etc.), and Central Asian states are not willing to intensify their multilateral cooperation on transnational water resources, the farming sector will eventually come to a standstill, or will even go backwards, and its productivity levels will be maintained only by recruiting more women and children as a slave-like workforce. If there is not more investment in the improvement of farming techniques, including water management, from all points of view (political, technical, administrative), or some alternative economic activities are not implemented, unemployment will grow, thus strengthening other "niches of alternative employment", such as the informal economy, migration, drugs, smuggling, prostitution, etc. And women and children in farming exploitations will keep on working under very poor conditions, with no chance of receiving an education or anticipating a better future. Attempts at multilateral cooperation (Eurasian Economic Community, Economic Cooperation Organization, Central Asia Regional Economic Cooperation Program, for example) in economic matters have not succeeded in including water and its management in their programs of cross-border cooperation, basically because of the antagonism between the countries located on the headwaters of the rivers (Tajikistan, Kyrgyzstan) and those on the lower courses (Turkmenistan, Uzbekistan and southern Kazakhstan) about the construction of new Tajik and Kyrgyz dams.

In the sociocultural sphere, all the processes of national construction initiated at the beginning of the 1990s had in common the monopolization of the politic power by those ethnic groups considered as "title holders" during the Soviet era. These ethnic groups are today seen as the exclusive heirs to the borders, territories and infrastructures designed and built by the Soviets, while the "non-title-holder" ethnic groups feel marginalized, often stigmatized or second-class citizens. The constitution of the new Central Asian states is not based on an international consensus, but on the dominance of one ethnic group over the rest. The dominant group, whose current legitimacy comes from their own consideration as heirs to the old Soviet legality—that is, from the exclusive founders of the current states, as well as from comprising the great majority of the population in an artificially delimited territory. This situation causes much internal tension among the different nationalities settled in the five Soviet republics, and Central Asian governments have in some cases reacted with attempts at manipulation and control over protest associations or organizations, as well as direct repression with no respect for human rights in other cases.

From a political perspective, the main threat to human security in the region comes from the precariousness or fragility of some state institutions with regard to the interests and motivations some leaders and their patronage networks have, as well as the lack of commitment of their political cadres to international conventions regarding human rights and the human right to water. The main consequence of the institutions' instability and weakness is that the allocation of resources, in this case water, the most important economic driving force of the region, is carried out in a non-transparent way and favors particular interests at the expense of general ones and the human right to water. These situations often cause discriminatory practices, whose first victims are minority ethnic groups and other less influential groups or social strata. The arbitrary and non-transparent water resource allocation becomes an essential mechanism that allows for social and political control, through which both governments and elites exchange protection for political acquiescence. Water management in Central Asia has become a power resource for certain elites rather than becoming a welfare tool for a state's citizens, depriving the majority of them of the most valuable human right: the right to water and sanitation.

In this context of threats and responses, the role of multilateral security organizations has not been very admirable. These organizations have not considered, as a priority, coping with the extensive conflict resolution issues and the threats derived from an insufficient level of cooperation and water resource management Consequently they have not foreseen, despite the multidimensional idea of security that some of them have, multilateral preventive policies against transnational threats affecting the implementation of the HRWS, such as droughts, floods, pollution, border and ethnic troubles, global warming, disputes on the use of water, corruption, and nepotism in water resources allocation. The resolution of all these conflicts involves cooperative transnational solutions, as well as transparency and civil society participation policies in the allocation and consumption of water resources. The failure of these international organizations to achieve sufficient cooperation, and to soften tensions and conflicts between Central Asian countries, could be attributed to several factors, some of them related directly to the organizations themselves, and some not:

1. Central Asian countries see multilateral security organizations, as well as some initiatives for cooperation in the matter of water resources, as a context in which to obtain some profit and privilege, but without any commitment to those questions affecting what they consider to be their national interests.

2. The consideration of these organizations by their most relevant members (Russia, China, the EU, the USA, etc.) as contexts to prioritize and defend or promote their interests, instead of environments of cooperation among them.

3. The lack of participation of actors coming from the civil society in the decision-making process within international security organizations, with the goal of democratizing global governance.

4. The absence of Afghanistan from the organizations and cooperation initiatives. This country is usually marginalized in regional negotiations discussing water resources management. None of the multilateral security organizations examined has Afghanistan as a full member, even the Fund for Saving the despite it being a key actor in water resources management in the Amu Darya basin, because 27.5 % of this watercourse is generated in its territory and the country shares an extensive fluvial border with Uzbekistan, Tajikistan and Turkmenistan (Klemm 2010).

5. The preeminence of Central Asian states' national interests over international legislation in matters of water resources management. In Central Asia, the needs of ecological and political systems are in

conflict. The processes of the creation of states, like those of economic management models, have ignored ecosystems and historical patterns of human settlement in the establishment of political boundaries. In several cases, major river systems served as a convenient means of demarcating new states. In a number of other cases, the lines drawn on a map out of colonial political consideration split river basins between upstream and downstream countries. As a result of this, international river basins water resources management today is therefore often a matter of international politics.[5]

6. Regional and geopolitical rivalries among the five states of the area, especially between the upstream countries (Kyrgyzstan and Tajikistan) and the downstream ones (Kazakhstan, Uzbekistan and Turkmenistan).[6]

7. The Russian role as intermediary in the construction of new hydrologic projects and their energy interests in the region.[7]

HUMAN SECURITY, HRWS AND EXTRAORDINARY POLITICS: TOWARDS A COMMUNITY OF SECURITY

The reinforcement of human security, empowerment and emancipation of citizens in this region of the world is indissolubly dependent on a new management of water based on the principles of the human right to water and sanitation, because the water-related policies carried out to date, begun in the Soviet era and based on an economist concept of water, have failed to guarantee the human right to water and sanitation for millions of people, as well as more recent water policies based on the particular interests of each state, which had led to a context of a "cold war of water" among the Amu Darya riverside countries. We need to raise water policies to a category of "extraordinary politics". Extraordinary democratic politics refers to those infrequent and unusual moments when the citizenry, overflowing the formal borders of institutionalized politics, aims reflectively at the modification of the central political, symbolic and constitutional principles, and at the redefinition of the content and purposes of a community (Kalyvas 2008, p. 7). Consequently, the principles that should govern the relationships between the different actors involved in the consumption and exploitation of water must be those of the HWRS.

The HWRS has several principles that offer the possibility of securitization of water resources as an open and democratic process, overcoming those contexts of exceptional politics that do not involve the empowering of individuals and their communities with regard to water resource management.

The main guidelines of HRWS are:

1. The human right to water and sanitation; this consists in considering that every person has the right to a sufficient amount of healthy water for personal and domestic use, as well as the right of access to sanitation.
2. Realization by states of the right to water and sanitation; all the spheres of a state's government, including the national government, regional administrations and local authorities, must progress gradually, but as fast as possible, towards the full individual exercising of the human right to water and sanitation, through a specific and well-oriented management, and must also take full advantage of all the available resources.
3. Preventing discriminatory measures and fulfilling all the needs of vulnerable or marginalized groups; the states must ensure that no individual or public or private organization applies discriminatory measures that could affect access to water and sanitation because of sex, age, ethnic origins, language, religion, political or any other kind of opinions, national or social background, disabilities, health condition or anything similar. The states must pay special attention to the needs of people and groups who are vulnerable, and traditionally have obstacles in the way of exercising their right to water and sanitation: women, children, indigenous peoples, people living in disadvantaged rural and urban areas, nomadic communities, etc.
4. The availability and equitable allocation of water; essential uses of water, personal and domestic, must have priority in terms of supply for all the people. To exercise their right to adequate food and to earn their living by means of work, marginalized or underprivileged farmers, as well as the remaining vulnerable groups, should have priority in matters of access to water resources to fulfill their basic needs. The right to water must be exercised in conditions of sustainability for current generations and those to come.
5. Improving access to water; the states must ensure that all the people have access to services of water supply and sanitation; the distribution of these services must be equitably guaranteed. If there

are not enough resources to guarantee high quality services, the states must above all invest in services that prioritize the needs of people with no basic access, usually by means of inexpensive services that can be improved, rather than expensive ones destined only to benefit a small section of the population.

6. Affordable price; the states must ensure that the price policies of water and sanitation are suitable, particularly foreseeing flexible payment modalities and subventions through which high-income users can help low-income ones.

7. Quality of water; the states should draw up water quality rules based on the World Health Organization's technical guidelines and the needs of vulnerable groups, taking the users' views into account. The states should also design regulations and policies to control water resource pollution coming from the populace and public or private organizations, foreseeing surveillance activities, disincentives and fines in cases of pollution, as well as some kind of encouragement to respect the rules.

8. Rights of participation; every person has the right to participate in the decision-making process when it affects their right to water and sanitation. It is especially necessary to guarantee that vulnerable or traditionally marginalized groups, especially women, are equitably represented in the decision-making process.

9. Resources and surveillance; each person should have access to administrative or judicial sources to report actions or omissions against the right to water and sanitation carried out by individuals and public or private organizations. The states must ensure the fulfillment of duties related to the rights to water and sanitation; for example, by creating or authorizing independent institutions, such as human rights commissions or regulating bodies, to be in charge of totally transparent surveillance and to be seen by users to take responsibility.

10. International responsibility and the duty of solidarity; the states should abstain from applying measures that could be an obstacle to exercising the rights to water and sanitation of people in other countries; they must also prevent other individuals and companies under their jurisdiction from doing the same. Bilateral and multilateral aid in the sector of water and sanitation should be oriented, as a priority, to the countries that are unable, by themselves, to enable their populations to enjoy the essential aspects of the rights to water and sanitation. International organizations, including

specialized bodies of the United Nations, especially those in the financial sector, and the member states of these organizations, should ensure that the rights to water and sanitation are respected in their policies and actions. The states should consider rights to water and sanitation when they design and apply international agreements with possible repercussions for that right.

All these and other principles must be included in a possible Amu Darya Security Community in order to guarantee the human rights of its populations, as well as the resolution of conflicts related to water.

TOWARD A SECURITY COMMUNITY IN THE AMU DARYA

To date, neither cooperation initiatives nor international organizations have achieved, in spite of their efforts, a sufficient degree of cooperation among Central Asian countries around the cross-border course of the Amu Darya, for reasons I have shown throughout this research. There are enough cooperation scenarios (initiatives and organizations) of different political, economic and environmental natures in the region— perhaps too many— and all have failed in one way or another to provide Central Asian states with common values about water because they have not taken that possibility into account. Deutsh says that "a community consists of persons who have learnt to communicate and understand each other beyond the simple exchange of goods and services" (Adler and Barnett 1998). Community derives from shared understandings of common interests and common identity among a group of people; without a shared identity, a sense of "we-ness," there is no community.[8]

In this sense, we discuss "community not as a matter of feelings, emotions, and affection, but as a cognitive process through which common identities are created." Human communities are therefore "imagined communities," socially constructed and intersubjectively understood by their members. Communal identity, moreover, is an essential component of broad human security; although a critic of a human security approach, Barry Buzan, has observed that, as social creatures, "individuals are not free standing, but only take their meaning from the societies in which they operate" (Schnurr and Swatuk 2012).[9]

States and organizations, apart from widening and empowering their political, economic or cultural exchanges, must create a community language that leads, through social learning, to a system of collective beliefs and perceptions with the aim of creating a common identity to promote mutual trust and to enable all kinds of peaceful conflict resolution.[10]

Therefore we strongly believe it is necessary to create a new water culture based on the HRWS criteria, with this element as a welfare resource to allow the existence of some expectations of pacific and welfare relations; if this product is only conceived, on the contrary, as an economic good, it will create inequalities difficult to be assumed, and if it is seen as a power resource it will cause uncertainty and conflict. The most important issue is to create a new culture of water based on the HRWS guidelines to replace the traditional one that, linked to national interests, still prevails in Central Asia, because it is not valid to manage a resource adequately that does not respect borders or ideologies, an idea that involves strengthening the role of both non-governmental organizations and users' associations. Locating water resources management within the state-centric discourse of "security"—of war, peace and the control of resources—is, in our estimation, a fundamental mistake (Swatuk and Vale 2001).

Adler and Barnett state that a positive and dynamic relation between structures of power and knowledge, on the one hand, and transactions, institutions, organizations and forms of social learning, on the other, generate two basic elements: trust and identity. Both aspects, reciprocally and mutually strengthening, enable the birth of a "Security Community" (Adler and Barnett 1998). As a consequence, the region's multilateral security organizations should promote practices and values to build up identity and trust around the management of water resources, leaving aside some security concepts as the dilemmas currently controlling the relations between Central Asian states in the matter of water resources, and replacing these values by those of the collective security community.

The creation of a security community focused on the HRWS could be an answer to the current lack of cooperation on water resources in Central Asia, showing the advantages of this resource when it is shared and not hoarded (zero-sum games) though this new culture of water, based on the HRWS, would obviously demand, a learning process, a process of construction of a new common identity built around an indispensable element for human life and the welfare of people in Central Asia: water.

To introduce these new ideas and values, and to be able to build this community of security around water, we need actors willing to carry out changes in the structures and values of the international security organizations, such as the Transnational Advocacy Networks (TANs), whose purpose is to defend the human right to water and sanitation.

To conclude, the construction of a security community in Central Asia demands a broad participation not only of formal and informal actors, and at every level (local, national, international), but it must also be a participation with a common purpose: to guarantee the human right to water and sanitation as the cornerstone for the rest of human rights and the process of emancipation of its citizens.

NOTES

1. Regional Migration Report: Russia and Central Asia. Edited by: Anna Di Bartolomeo, Shushanik Makaryan and Agnieszka Weinar. This Report has been published by the European University Institute, Robert Schuman Centre for Advanced Studies, Migration Policy Centre within the framework of the CARIM-East project. 2014, p. 15.
2. World bank Europe and Central Asia. http://www.worldbank.org/en/publication/global-economic-prospects/regional-outlooks/eca.   Europe and Central Asia.
3. Impact of Climate Change to Water Resource in Central Asia, EDB and EB. IFAS, 2009.
4. *Global Water Security*, Defense Intelligence Agency, 2 February 2012.
5. Security, Ecology Community: Contesting the Water Wars. Hypothesis in Southern Africa. L. A. Swatuk, L. Thompson, M. Hara, P. Van Der Zaag.
6. Water Pressures in Central Asia Europe and Central Asia. Report No. 23311. Sept 2014.
7. Pérez Martín, M. A. (2009): *Geo-Economics in Central Asia and the "Great Game" of Natural Resources: Water, Oil, Gas, Uranium and Transportation Corridors*. WP 59/2009 (Translated from Spanish), 19 April 2010. http://www.realinstitutoelcano.org/wps/portal/web/rielcano_en/contenido?WCM_GLOBAL_CONTEXT=/elcano/elcano_in/zonas_in/dt59-2009#.VNn3lhs5DIU.
8. Anderson, Benedict (2006): *Imagined Communities: Reflections on the Origin and Spread of Nationalism*. London: Verso.
9. Adler, Emanuel (2005): *Imagined (Security) Communities: Cognitive Regions in International Relations*, in Emanuel Adler (ed.), Communitarian International Relations: The Epistemic Foundations of International Relations (New York: Routledge), p. 195.
10. Buzan, Barry (2004): *A Reductionist, Idealistic Notion that Adds Little Analytical Value*, Security Dialogue, Vol. 35, No. 3, p. 370.

REFERENCES

Adler, E., Barnett, M. 1998. *Security Communities.* Cambridge University Press. p. 6–10.

Alkire, S. (2003). A conceptual framework for human security. CRISE. Working Paper No. 2, Queen Elizabeth House, University of Oxford, 2003.

Aquastat survey. 2012. *Irrigation in central Asia.* Edited by FAO, p. 67.

Allouche, J. (2007). The governance of Central Asian waters: National interest versus regional cooperation. *Central Asia at the Crossroads.* www.unidir.org/pdf/articles/pdf-art2687.pdf.

Avoiding Water Wars: Water Scarcity and Central Asia's Growing Importance for Stability in Afghanistan and Pakistan. (2011). A majority staff report prepared for the use of the committee on Foreign Relations United State Senate one hundred twelfth congress. First Session. February 22, pp. 1–10.

Bacquelaine, D. (2003). Sub-committee on NATO partnerships (p. 7). http://www.nato-pa.int. June 2013.

Bakker, K. J. (2003). Uncooperative commodity privatizing water in England and Wales. Geographical and Environmental Studies Series. New York: Oxford University Press.

Balzacq, T. (2008, January). The policy tools of securitization: Information exchange, EU foreign and interior policies. *JCMS Journal of Common Market Studies, 46*(1), 75–100.

Balzacq, T. (2010). *Securitization theory: How security problems emerge and dissolve.* London: Taylor & Francis.

Banks, A. S., Day, A. J., Muller, Th. C., 0, 0. (1998). *Political handbook of the world 1998* (p. 1003). London: Palgrave MacMillan.

M.Á. Pérez Martín, *Security and Human Right to Water in Central Asia*, Security, Development and Human Rights in East Asia, DOI 10.1057/978-1-137-54005-8

Berg-Schlosser, D. (2007). *Democartization: The State of the Art, The World of Political Science-the Development of the Discipline, Volume 1 of World of political science,* edited by Michael Stein and John Trent. Opladen and Farmington Hills: Barbara Budrich Publishers.

Bernstein, S. (2001). The compromise of liberal environmentalism. New York: Columbia University Press.

Booth, K. (1991). Security and emancipation. *Review of International Studies, 17* (4), 313–326. 26 Ibidem, p. 319.

Brown, L. (1977). Redefining national security, Worldwatch Institute paper #14. Washington, DC: Worldwatch Institute.

Buzan, B., Waever, O., and de Wilde, J. (1998). *Security: A new framework for analysis.* Boulder, CO: Lynne Rienner Publishers.

Christie, R. (2010). Critical voices and human security: To endure, to engage, or to critique? *Security Dialogue, 41,* 169–190.

Council of the European Union General Secretariat. 2007. *European Union and Central Asia: Strategy for a New Partnership.* http://trade.ec.europa.eu/doc lib/docs/2008/october/tradoc_141165.pdf

Cummings, S. N. (Eds.) (2002). *Power and Change in Central Asia Politics in Asia.* Series. Edition, illustrated. London: Routledge.

EDB and EB IFAS. 2009. Eurasian development bank executive board of the international fund for saving the Aral sea regional center of hydrogeology. *Impact of Climate Change to Water Resources in Central Asia,* 10–22.

Falkenmark, M., & Lindh, G. (1974). Impact of water resources on population. Swedish delegation to the UN world population conference, Bucharest.

Giordano, M. A., & Wolf, A. T. (2003). Sharing waters: Post-Rio international water management. *Natural Resources Forum, 27,* 163–171.

Garcés de los Fayos, F. (2014). In-depth analysis. The World Bank considers feasible the building of the Tajik Rogun dam. http://www.europarl.europa.eu/ RegData/etudes/IDAN/2014/536392/EXPO_IDA(2014)536392_EN.pdf

Giffen, J., & Earle, L. with Charles Buxton. (2005). *The development of civil Society in Central Asia.* INTRAC's Publications Central Asia Programme. https:// assets.publishing.service.gov.uk/media/57a08c3de5274a27b200108f/ R7649-report.pdf

Gleick, P. H. (1992). Environmental consequences of hydroelectric development: The role of facility size and type. *Energy: The International Journal, 17*(8), 735–747. Pergamon Press, Ltd., Great Britain.

Gleick, P. H. (1993). Water and conflict. *International Security, 18*(1), Summer, 79–112.

Goldman, M. (2005). Imperial nature: The world bank and struggles for social justice in the age of globalization. New Haven, CT: Yale University Press.

Homer-Dixon, T. (1991). On the threshold: Environmental changes as causes of acute conflict. *International Security, 16,* 76–116.

Human Development Report. 2006. *Beyond scarcity: Power, poverty and the global water crisis.* New York: United Nations Development Programme (UNDP).

Human right Watch Web. (2014). "We Suffered When We Came Here," Rights Violations Linked to Resettlements for Tajikistan's Rogun Dam. June 25, 2014. https://www.hrw.org/report/2014/06/25/we-suffered-when-we-came-here/rights-violations-linked-resettlementstajikistans

IFAS *Aral Sea Basin Program 3 Document.* Serving the people of Central Asia. http://www.ec-ifas.org/engine/download.php?id=16

Kalyvas, A. (2008). *Democracy and the politics of the extraordinary. Max weber, carl schmitt, and hannah arendt* (p. 7). London: Cambridge University Press.

Kazantsev, A. (2008). Russian policy in Central Asia and Caspian sea region. *Europe-Asia studies, 20* (6), 1073–1088.

Keck, M. E., & Sikkink, K. (1998). *Activists beyond borders.* Ithaca: Cornell University Press.

Klemm, W. (2010). Sr. Land & water development engineer investment centre division: The Afghan part of Amu Darya basin. Impact of irrigation in Northern Afghanistan on water use in the Amu Darya basin. *FAO,* Rome and Sayed Sharif Shobair, Chief engineer and coordinator of the emergency irrigation Rehabilitation project of the ministry of energy and water of Afghanistan, FAO Kabul http://www.unece.org/fileadmin/DAM/SPECA/documents/ecf/2010/FAO_report_e.pdf

Kranz, N., Vorwerk, A., & Interwies, E. (2005). Transboundary river basin management regimes: The Amu Darya basin case study. First draft completed July 19, 2005. *RBA* July 22, 2005.

Krause, K., & Williams, M. C. (1997). From strategy to security: Foundations of critical security studies. In K. Krause & M.C. Williams (Eds.), *Critical security studies: Concepts and cases* (p. 43). Minneapolis, MN: University of Minnesota Press.

Lipschutz, R. (1995). *On security* (p. 10). New York: Columbia University Press.

Mayer, P. (2008). Civil society participation in international security organizations: The cases of NATO and the OSCE. In J. Steffek, C. Kissling, & P. Nanz (Eds.), *Civil society participation in European and global governance: A cure for the democratic deficit? (Transformations of the State)* (p. 116). Basingstoke: Palgrave Macmillan.

Micklin, P. P. (2000). *Managing water in Central Asia. Central Asian and Caucasian prospects* (p. 231). London: The Royal Institute of International Affairs.

Molchanov, M. A. 2012. *Eurasian Regionalisms and Russian Foreign Policy. A paper prepared for the UACES Convention Exchanging Ideas on Europe 2012,* Passau, Germany, 3–5 September 2012. http://uaces.org/documents/papers/1201/molchanov.pdf.

Nef, J. (1999). *Security and mutual vulnerability. The international political economy of development and underdevelopment.* Ottawa: IDRC Books.

North Atlantic Treaty Organisation. 2014. Office of the nato liaison officer (nlo) in central Asia completed nato programmes in Central Asia Environment and Security (ENVSEC) Initiative. http://www.nato.int/cps/en/natolive/109965.htm.

Organization for Security and Co-operation in Europe. (2010, December 3). Astana Commemorative Declaration Towards a Security Community (p. 3), Astana. Summit Meeting Astana 2010 Second day of the Astana Summit Meeting. SUM(10) Journal No. 2, Agenda item 4.

Pérez Martín, M. A. (2009). La geoeconomía de Asia Central y el "Gran Juego" de los recursos naturales: Agua, petróleo, gas, uranio y corredores de transporte (DT). *Real Instituto Elcano*. DT 59/2009 – 23/11/2009. http://www.realin stitutoelcano.org/wps/portal/rielcano/contenido?WCM_GLOBAL_ CONTEXT=/elcano/elcano_es/zonas_es/asia-pacifico/dt59-2009.

Rogers, P. (2002). Water is an economic good: How to use prices to promote equity, efficiency, and sustainability. *Water Policy, 4*(2002), 1–17.

Schmitt, C. (2005). *Political theology: Four chapters on the concept of sovereignty* trans. by George Schwab. Chicago: University of Chicago Press.

Schmitz, H. P., & Sikkink, K. (2002). *Handbook of international relations* (pp. 827–851). London: Sage.

Schnurr, M., and Swatuk, L. (Eds.) (2012). Insecurities of non-dominance: Re-theorizing human security and environmental change in developed states. In *Natural resources and social conflict towards critical environmental security*. Wilfrid Greaves, p. 68.

Serving the People of Central Asia. (2011). About EC IFAS. http://ec-ifas.water unites-ca.org/about/index.html

Spoor, M. (1993). Transition to market economies in former Soviet Central Asia: Dependency, cotton, and water. *The European Journal of Development Research, 5* (2), 142–158.

Swatuk, L. and Vale, P. 2001. Why democracy is not enough: Southern Africa and human security. In N. Poku (Ed.), *Security and development in southern africa*. Westport: Praeger.

Ullman, R. H. (1983). Redefining security. *International Security*, 8, 129–153.

UNEP_GRIDA Environment and Security in the Amu Darya River Basin. (2011). http://www.grida.no/publications/security/book/4881.aspX

United Nations. (2010). *International decade for action 'water for life' 2005–2015*. Resolution 64/292. New York: United Nations.

United Nations Environment Programme (UNEP). 2014. Thematic focus: Environmental governance, Ecosystem management, Climate change The future of the Aral Sea lies in transboundary co-operation, pp. 2–5. https:// na.unep.net/geas/archive/pdfs/GEAS_Jan2014_Aral_Sea.pdf.

Veldwisch, G. J. A. and Spoor, M. (2008). Contesting rural resources: Emerging 'forms' of agrarian production in Uzbekistan. *Journal of Peasant Studies* 35(3): 424–451.

Waever, O. (1995). Identity, integration and security: solving the sovereignty puzzle in EU studies. *Journal of International Affairs, 48*, 389–431.

Wegerich, K. (2002). Natural drought or human made water scarcity in Uzbekistan? *Central Asia and the Caucasus, 2*(14), 154–162.

Wendt, A. (1992). Anarchy is what states make of it: The social construction of power politics. *International Organization, 46*(2), Spring, 391–425.

Wendt, A. (1999). *Social theory of international politics.* Cambridge University Press, ISBN 0-521-46960-0.

Westing, A. H., ed. (1986). An expanded concept of international security. In *Global resources and international conflict.* Oxford: Oxford University Press.

Wiegmann, G. (2011, April). *The Role of Local Institutions in the Statehood-Building Process in Tajikistan.* Paper, CARN-conference "Dynamics of Transformation in Central Asia – Perspectives from the Field", Rome.

Williams, M. C. (2015). Securitization as political theory: The politics of the extraordinary. *International Relations, 29*(1), 114–120.

Winpenny, J. T. (1994). *Managing water as an economic resource.* London/New York: Routledge.

Wolf, A. T. (1999). Water and human security (prepared for the Global Environmental Change and Human Security Project).

INDEX

© The Author(s) 2017   173
M.Á. Pérez Martín, *Security and Human Right to Water in Central
Asia*, Security, Development and Human Rights in East Asia,
DOI 10.1057/978-1-137-54005-8